Basic Principles of

Agricultural Meteorology

Basic Principles of

Agricultural Meteorology

Dr. V. Radha Krishna Murthy
M.Sc. (Ag.) Ph.D.
Associate Professor, Department of Agronomy,
Acharya N.G. Ranga Agricultural University,
College of Agriculture,
Rajendranagar, Hyderabad, A.P.,
INDIA - 500 030.

4-4-309/316, Giriraj Lane, Sultan Bazar,
Hyderabad - 500 095
Phone : 040 - 23445605, 23445688

Published by :

A unit of **BSP Books Pvt., Ltd.**

4-4-309/316, Giriraj Lane, Sultan Bazar,
Hyderabad - 500 095
Phone : 040 - 23445605, 23445688
e-mail : info@bspbooks.net

ISBN : 978-93-52300-74-7 (HB)

Dr. I.V. SUBBA RAO
VICE- CHANCELLOR

ACHARYA N.G. RANGA AGRICULTURAL UNIVERSITY
Rajendranagar, Hyderabad-500 030
Phones : Offi. : 4015035
Resi.: 7752233, 7750631
Grams : "AGRIVARSITY"
Fax : 91-040-4015031
E-mail : Vc_angrau@yahoo.com
root.@apau.ren.nic.in

FOREWORD

All the World over, agriculture has cradled culture and civilization. It is the mother of all industries and maintainer of life on earth. It holds key for the socio-economic development and political stability of any nation. The nature is the natural resource for agriculture. Man has been practicing farming, since ages exploiting nature, for his survival and prosperity. With the growing population and its needs, nature was subjected to over exploitation at the expense of sustainability of resources and production base.

Resources and efforts alone do not lead to increased production. They have to be supported by nature, in the form of favourable weather and monsoons, crucial to agricultural production. We have succeeded in enhancing our agricultural production to steer the Nation from a 'food deficit' to 'food surplus' status, which became possible through the sustained efforts of scientists, farmers and policy makers, well supported by mother Nature by way of favourable monsoons. Our progress in this front, however, has been put to set back, at times, whenever the nature is adverse. With the advancements in Space Science, Information Technology, we are now in a better position to predict and forecast the behaviour of weather and monsoons reasonably well to safeguard ourselves, though not dictate their behaviour. Of the sciences that help us in understanding and adjusting with the Nature in relation to agriculture, Agricultural Meteorology is one. As expressed by the author in his book, it is still a young subject of study in India, not so familiar to the students of Agriculture. As such, they look for suitable study material on the basics of this subject. Though there are several advanced books on the subject, text books on basic principles of the subject are wanting. This has prompted Dr. Radhakrishna Murthy, to respond to this by focusing his efforts on this felt-need. Though basically an Agronomist, he seems to have developed a flair for Agrometeorology, as is evident from this venture, apart from his earlier publications on the same subject.

A cursory glance at the contents and the way of their presentation, reveal the urge of the author to help the readers to know and understand the basics of this less- known, but important, subject for agriculture, with all possible ease. I am sure this book will not only serve as a Text on the subject for the Teachers and Students, but as a source of information to all those interested and involved in the management of agriculture and allied sectors in relation to weather and its behaviour.

I compliment the efforts of Dr. Radhakrishna Murthy for his endeavour to bring out this useful publication. It is hoped that this book will be of use for students of Agricultural Meteorology and serves as a source material for their use.

Dated: 4th September, 2002

L. V. SUBBA RAO

Phones : Offi. : 4015226
4015161 Ext. 307
4015011 308
Resi : 7639734
Grams : "Agriversity"
Fax : 91-040-4015031
E-mail : root@apau.ren.nic.in

ACHARYA N.G. RANGA AGRICULTURAL UNIVERSITY
(Formerly Andhra Pradesh Agricultural University)
Administrative Office : Rajendranagar, Hyderabad-500 030.

Dr. M.V. SHANTARAM
DEAN
Faculty of Post-Graduate Studies (Retd.) &
COORDINATOR
SPECIAL PROGRAMMES

PREFACE

Agricultural Meteorology has developed into an independent field of study over the past five decades on its own merit. The students of earth and atmospheric sciences, more particularly those interested in agriculture, depend a lot on weather.

Globally, climate change is exerting an enormous influence on productivity of both natural and cultivated ecosystems. Climate change has now become a hot topic of discussion in several fora world wide. A clear and concise discussion on weather and related aspects is the urgent need of the hour. Such a compilation should be able to put forward basic concepts of weather related aspects and its relation to agriculture, crop productivity and primary productivity of natural ecosystems. It is preferable that the information on weather is conveyed in very simple terms to students of agriculture who are basically interested in applied aspects of meteorology. The author of this publication on "Agricultural Meteorology" has attempted to do this. The individual chapters deal with different components of weather. Following the basic discussion, Chapter 9, 10 and 11 deal with applications of Meteorological data for tackling serious problems of crop production. The discussion on Disaster Management, Weather Forecasting, Synoptic Reports and Remote Sensing are very useful for a student of Agriculture or a practical farmer. Likewise in Chapters 10 and 11 Crop Growth Modelling, Climate change, Micrometeorology and Weather Modification have been discussed which in the present day context of climate change which is affecting productivity of agriculture world wide is very appropriate. All in all the book on Agricultural Meteorology is a painstaking output by Dr. V. Radhakrishna Murthy who is by training an Agronomist with specialization in Agricultural Meteorology.

I am sure, that the publication will be useful to both students of Agriculture and Environmental Science as well as practising Agro-meteorologists and Progressive farmers.

Dated: 8th July, 2002

HYDERABAD

M.V.SHANTARAM

ACKNOWLEDGEMENTS

I prostrate with the highest devotion before the majestic, gorgeous, rapturous and resplendent presiding deity The Lord of Seven Hills, *Kaliyuga Daivam* Sri Tirumala Tirupati Srivenkateswara Swami for HIS copious blessings on me in accomplishing this work.

It gives me immense pleasure to express my deep sense of gratitude, indebtedness and sincere regards to ***Padma Shri*** **Dr. I. V. Subba Rao,** Hon'ble Vice - Chancellor of Acharya N. G. Ranga Agricultural University, Hyderabad, who has agreed to write the foreword to this book. His works are a part of history of Indian agriculture. Without his unstinted attention I might not have succeeded in completing this work.

I profusely thank Professor M. V. Shantaram, an eminent academician and innovative researcher in agricultural sciences for writing the preface. I am profoundly indebted to Dr. N. Sreerama Reddi, Dr. S. Raghu Vardhan Reddy and Dr. A. Padma Raju, the top dignitaries of the university for their unparalleled encouragement to me. I am thankful to Dr. M. V. K. Siva Kumar, Chief, Agricultural Meteorology, WMO, Geneva, Dr. T. Yellamanda Reddy, Dr. V. Satyanarayana, Dr. N. Venkat Reddy, Dr. A. S. Raju and Dr. B. Bucha Reddy of ANGRAU for their kind help in several ways in my professional and personal development. Their support to me is exemplary. Similarly, I express my ineffable gratitude to Professor Kees Stigter, Wageningen University, The Netherlands, Professor J. W. Jones, Professor K. J. Boote and Dr. P. V. V. Prasad of University of Florida, USA and Drs. A. Latchanna, Shaik Mohammed, N. G. Rao, Y. G. Rao, A. R. Rao, V. V. Ranga Rao and G. V. Peterson of ANGRAU for sparing their valuable and precious time in giving several suggestions for improvement of this text. I salute all these stalwarts with reverence. May the divine grace be with them.

I pay my highest respects from the inner core of my heart to my late parents and always pray with folded hands for their many sacrifices in bringing me to this level and for their unending divine source of affection and moral strength. The unparalleled support and affection in many ways by both my in-laws and all my other family members are gratefully acknowledged.

Sri M. Raja Mohan Reddy, a doyen of peasants and formerly Member of Parliament has been a source of great strength to all our families. It is a great honour to me to acknowledge his help to all of us on many an occasion.

I am greatly beholden and owe a deep sense of honour to my eldest brother Sri V. L. Kantha Rao whose moral support in shaping my career will go a long way throughout my life.

I feel it great privilege to acknowledge with immense pleasure the affection of my beloved colleagues and well wishers Sri J. Purnachandra Rao, Sri. M. Ravi Kumar and Sri V. M. Prasad.

My wife typed the manuscript and brought this book to the present shape as she did for all the previous documents brought out by me as well as the other books in preparation. Diction is not enough and words fall short to express my gratitude for her deep care and for her understanding the philanthropic principles which I follow in my life. The ever blooming and cheerful faces of my daughter Chy. V. A. Pooja and son Chy. Srivenkatesh have always been a tremendous relaxation to me. I love them a lot. I wish they choose some time in future to read this book.

The study material, lecture notes, figures (totally redrawn with the latest software packages) tables, etc., in this book are compiled, scientifically reviewed and narrated from several books that I have read over the years. While I have taken care to establish the identity (mentioned in the references section) of the author of each small item, diagram, matter, etc., this has always not been possible. I regret any inadvertent error or omission and would appreciate it being brought to my notice. I acknowledge their works with utmost respect and dignity.

I am deeply grateful to Mr. Nikhil Shah and Anil Shah for making my dream a reality by bringing this book through the BS Publishers. I have many words of appreciation for the services extended to me by Mr. Naresh who had given excellent shape through his art to this book.

The mantras from "SRI ISOPANISAD" by His Divine Grace A. C. Bhaktivedanta Swami Prabhupada are embedded at the beginning of each chapter. These were given to me by several noble hearted souls at SRI KRISHNA TEMPLE, Gainesville, USA. I hope that these invaluable and most powerful mantras definitely give moral strength to the readers to perform the best in all walks of their lives and for material well being with them.

I wish that this work from me today must be helpful for the food security of the poorest of the poor of tomorrow.

Author.

My Beloved DEAN, Faculty of Agriculture, ANGRAU, Hyderabad.

DR. N. SREERAMA REDDI

Dr.N. Sreerama Reddi has an excellent Academic and Scientific background. He got an exemplary career which inspires to be emulated by one and all. He was born on 27 October 1942 at Zulakallu Village of Guntur (District), A.P. India. After Bachelors and Masters Degree programmes in India he did his Ph.D from Kansas State University, USA and IARI, New Delhi. He is an eminent Scientist in Genetics and Plant Breeding with special reference to Hybrid Rice Breeding and published over 50 Scientific articles. He has a wonderful teaching, research and extension work experience in Agricultural Sciences and is responsible for the development, popularisation and release of genetically BPH resistant varieties in rice. His commitment, hard work and sincerity enabled him to shoulder highly coveted positions like Principal Scientist (Rice), Associate Dean and Dean of Agriculture in ANGRAU. He visited several countries (U.K, Australia, Philippines, etc.,) and attended and conducted several prestigious scientific and academic programmes. The seed village concept, students counselling, campus interviews, establishment of scientific clubs, committees, conducting high level NCC and sports camps, cultural programmes, etc., are his other notable achievements. The Author is proud to be a student of this eminent personality and salutes him with utmost respect.

Author.

Dedicated to

Lord Sri Tirumala Tirupati Venkateswara

CONTENTS

Chapter - 3

Chapter - 4

Chapter - 7

Chapter - 8

List of Tables

List of Figures

Chapter - 1

Atmosphere and Agricultural Meteorology

"Everything animate or inanimate that is within the universe is controlled and owned by the Lord. One should therefore, accept only those things necessary for himself, which are set aside as his quota, and one should not accept other things, knowing well to whom they belong".

ATMOSPHERE

The earth is elliptical in shape. It has three spheres. They are :

1. Hydrosphere : the water portion,
2. Lithosphere : the solid portion, and
3. Atmosphere : the gaseous portion.

The atmosphere is defined as, "The colourless, odourless and tasteless physical mixture of gases which surrounds the earth on all sides". It is mobile, compressible and expansible.

Uses of the Atmosphere for Agriculture

The uses of atmosphere are : It :

1. Provides oxygen which is useful for respiration in crops.
2. Provides carbon-dioxide to build biomass in photosynthesis.
3. Provides nitrogen which is essential for plant growth.
4. Acts as a medium for transportation of pollen.
5. Protects crop plants on earth from harmful U.V. rays.
6. Maintains warmth to plant life and
7. Provides rain to field crops as it is a source of water vapour, clouds, etc.

Composition of the Atmosphere

There is no definite upper layer to the atmosphere. The decrease of air (density) with altitude (height) is so rapid (Figure 1.1) that half of the atmosphere lies within 3.5 miles (5.5 kilometres) from the surface and nearly three fourth of the atmosphere lies upto 7 miles (11 kilometres).

The atmosphere is a mixture of many gases. In addition, it contains large quantities of solid and liquid particles collectively called "Aerosols". The lower part of the atmosphere contains water vapour from 0.02 to 4 per cent by volume. Nitrogen and oxygen make - up approximately to 99 per cent and the remaining 1 per cent by other gases (Table 1.1). Innumerable dust particles are also present in the lower layers of the atmosphere. They are microscopic and play an important role in absorption and scattering of insolation.

TABLE 1.1

PRINCIPAL GASES COMPRISING DRY AIR IN THE LOWER ATMOSPHERE

S. No.	Constituent	Per cent by volume	Per cent by weight
1.	Nitrogen	78.08	75.51
2.	Oxygen	20.94	23.15
3.	Argon	0.93	1.28
4.	Carbon-dioxide	0.03	0.046

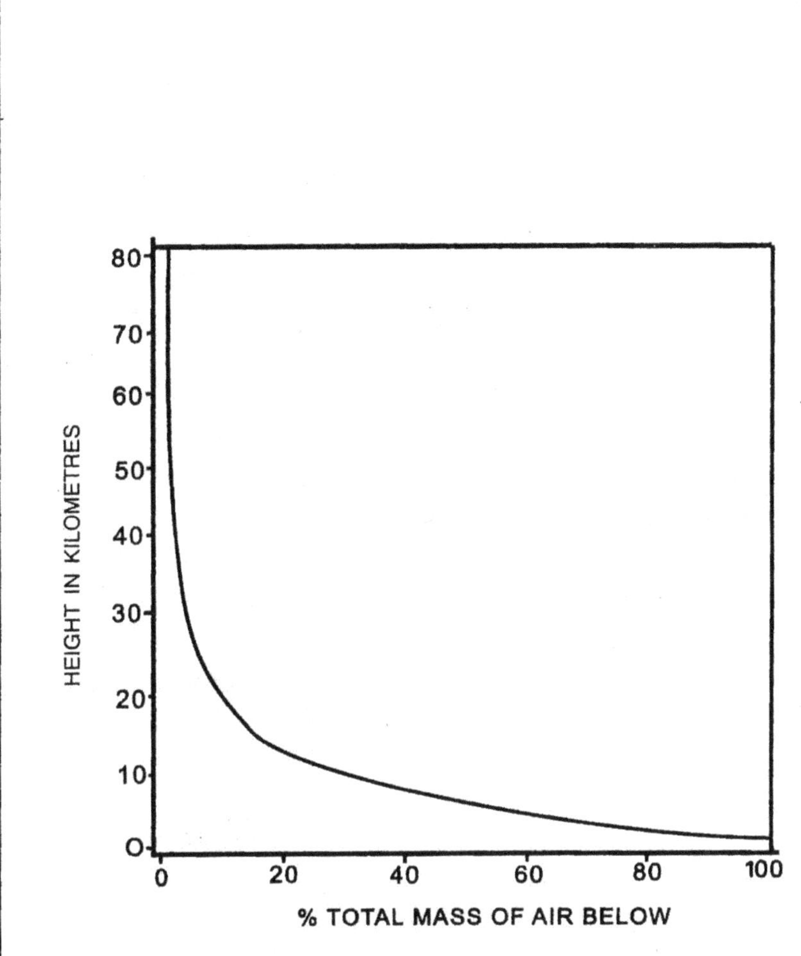

Figure : 1.1 Mass of the Earth's Lower Atmosphere

Physical Structure of the Atmosphere

On the basis of vertical temperature variation (Figure 1.2), the atmosphere is divided into different spheres or layers.

I. Troposphere

1. The word "Tropo" means mixing or turbulence and "Sphere" means region.
2. The average height of this lower most layer of the atmosphere is about 14 kilometres above the mean sea level; at the equator it is 16 kilometres; and 7- 8 kilometres at the poles.
3. Under normal conditions the height of the troposphere changes from place to place and season to season.
4. Various types of clouds, thunderstorms, cyclones and anti-cyclones occur in this sphere because of the concentration of almost all the water vapour and aerosols in it. So, this layer is called as "Seat of weather phenomena".
5. The wind velocities increase with height and attain the maximum at the top of this layer.
6. Another striking feature of the troposphere is that there is a decrease of temperature with increasing elevation at a mean lapse rate of about 6.5°C per kilometre or 3.6°F per 1,000 feet.
7. Most of the radiation received from the sun is absorbed by the earth's surface. So, the troposphere is heated from below.
8. In this layer, about 75 per cent of total gases and most of the moisture and dust particles present.
9. At the top of the troposphere there is a shallow layer separating it from the stratosphere which is known as the "Tropopause".
10. The tropopause layer is thin and its height changes according to the latitudes and it is a transitional zone and distinctly characterised by no major movement of air.

II. Stratosphere

1. This layer exists above the tropopause (around 20 kilometres onwards) and extends to altitudes of about 50-55 kilometres.
2. This layer is called as "Seat of photochemical reactions".

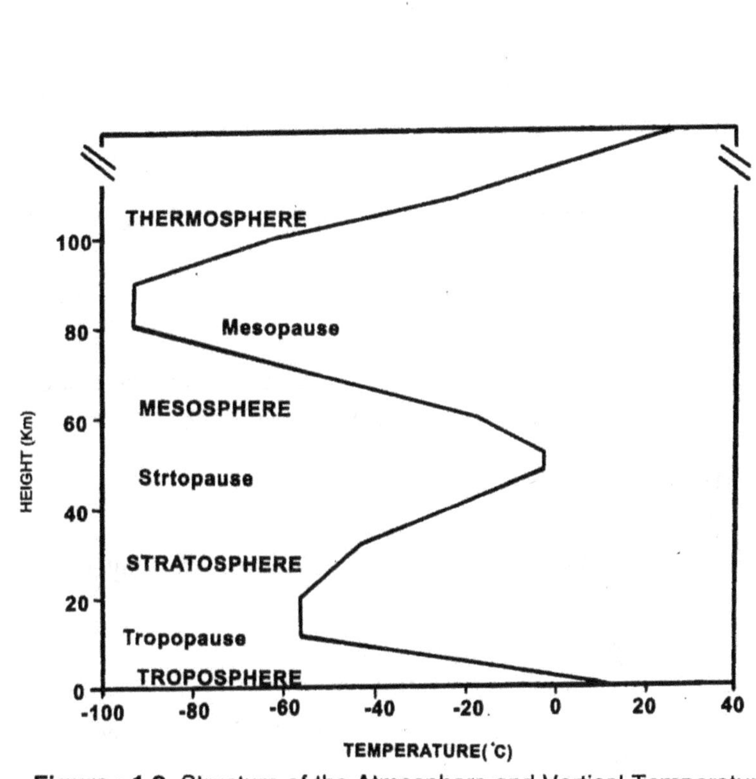

Figure : 1.2 Structure of the Atmosphere and Vertical Temperature Distribution

3. In any particular locality, the temperature remains practically constant at around 20 kilometres and is characterised as isothermal because the air is thin, clear, cold and dry.
4. The temperature of this layer increases with height and also depends upon the troposphere because the troposphere is higher at the equator than at the poles.
5. In the upper parts of the stratosphere the temperatures are almost as high as those near the earth's surface, which is due to the fact that the ultra - violet radiation from the sun is absorbed by ozone in this region.
6. Less convection takes place in the stratosphere because it is warm at the top and cold at the bottom.
7. There is also persistence of circulation patterns and high wind speeds.
8. The upper boundary of the stratosphere is called the stratopause and above this level there is a steep rise in temperature.

III. Mesosphere / Ozonosphere

1. There is a maximum concentration of ozone between 30 and 60 kilometres above the surface of the earth and this layer is known as the ozonosphere.
2. A property of the ozone is that it absorbs ultra violet rays. Had there been no layer of the ozone in the atmosphere, the ultra - violet rays would have reached the surface of the earth and no life on it.
3. The temperature of the ozonosphere is high (warm) due to selective absorption of ultra - violet radiation by ozone.
4. Because of the preponderance of chemical processes this sphere is called as the "Chemosphere".
5. In this layer the temperature increases with height at the rate of 5°C per kilometre.
6. According to some leading scientists the ionosphere is supposed to start at a height of 80 kilometres above the earth's surface. The layer between 50 and 80 kilometres is called as "Mesosphere". In this layer the temperature decreases with height. The upper boundary of this layer is called the mesopause.

IV. Ionosphere/Thermosphere

1. The Ionosphere layer lies beyond the ozonosphere (mesosphere) at a height of about 80 kilometres above the earth's surface and extends upto 400 kilometres.
2. The atmosphere in the ionosphere is partly ionised. Enriched ion zones exist in the form of distinct ionised layers. So, this layer is called as the ionosphere.
3. Above the ozonosphere the temperature falls again. According to some climatologists, the layer between 80 and 140 kilometres is known as the "Thermosphere".
4. The ionosphere reflects the radio waves because of one or multiple reflections of shortwave radio beams from the ionised shells. So, long distance radio communication is possible due to this layer .

V. Exosphere

1. The outer most layer of the earth's atmosphere is named as the exosphere and this layer lies between 400 and 1,000 kilometres.
2. At such a greater height the density of atoms in the atmosphere is extremely low.
3. Hydrogen and Helium gases predominate in this outer most region.
4. At an altitude of about 500 to 600 kilometres the density of the atmosphere becomes so low that collisions between the neutral particles become extremely rare.

Weather and Climate

Weather

It is defined as :

1. "A state or condition of the atmosphere at a given place and at a given instant of time".
2. "The daily or short term variations of different conditions of lower air in terms of temperature, pressure, wind, rainfall, etc".

The aspects involved in weather include small areas and duration, expressed in numerical values, etc. The different weather elements are solar radiation, temperature, pressure, wind, humidity, rainfall, evaporation, etc. Weather is highly variable. It changes constantly sometimes from hour to hour and at other times from day to day.

Example : The air temperature of Rajendranagar on 20-01-2000 at 2.30 p.m. is 32°C.

Climate

It is defined as :

1. "The generalised weather or summation of weather conditions over a given region during comparatively longer period".
2. "The sum of all statistical information of a weather in a particular area during a specified interval of time, usually, a season or a year or even a decade".

The aspects involved are larger areas like a zone, a state, a country and is described by normals. The climatic elements are latitude, longitude, altitude, etc.

Example : In Andhra Pradesh the winter temperatures range from 15 to 29°C.

TABLE 1.2

DIFFERENCES BETWEEN WEATHER AND CLIMATE

S. No.	Weather	Climate
1.	A typical physical condition of the atmosphere.	Generalised condition of the atmosphere which represents and describes the characteristics of a region.
2.	Changes from place to place even in a small locality.	Different in different large regions.
3.	Changes according to time (every moment).	Change requires longer (years) time.
4.	Similar numerical values of weather of different places usually have same weather.	Similar numerical values of climate of different places usually have different climates.
5.	Crop growth, development and yield are decided by weather in a given season.	Selection of crops suitable for a place is decided based on climate of the region.
6.	Under abnormal weather conditions planners can adopt a short-term contingent planning.	Helps in long-term agricultural planning.

Environment

The environment is defined as, "A collective form embracing all the conditions in which organisms live".

Sub-Divisions of the Environment

The environment is sub-divided into two :

1. Physical or abiotic or non-living environment.

   Example : Solar radiation, temperature, pressure, humidity, wind, etc.

2. Biotic or living environment.

   Example : Predators, parasites, soil organisms, insects, pests, etc.

Both of these environments effect the distribution of organisms in different habitats. The environment can also be divided into :

1. Micro-environment which is the immediate surrounding of an organism.
2. Macro-environment which is the total of physical and biotic environments that surrounds the organism and its micro-environment.

Different Groups of Climate

For easy understanding of study of the climate, it is expressed in three different ways. They are :

1. **Micro-climate :** The climate of relatively smaller areas i.e., a cropped field, a class room, etc.
2. **Meso-climate :** The climate of a region or location like a valley, forest, etc. This climate is intermediate between micro and macro-climate.
3. **Macro-climate :** The climate global in nature i.e., a state, nation ,etc.

AGRICULTURAL METEOROLOGY

Meteorology

Meteorology is defined as :

1. "The science of atmosphere".
2. "A branch of physics of the earth dealing with physical processes in the atmosphere that produce weather".

Climatology

It is defined as, "The science dealing with the factors which determine and control the distribution of climate over the earth's surface". Different factors affecting the climate of a region are :

1. Latitude.
2. Altitude.
3. Land and water.
4. Winds and air masses .
5. Low and high pressure belts.
6. Mountain barriers.
7. Ocean currents.
8. Extent of forests, etc.

The above factors are also known as the "Climatic elements".

TABLE 1.3

DIFFERENCES BETWEEN METEOROLOGY AND CLIMATOLOGY

S. No.	Meteorology	Climatology
1.	This is derived from the Greek word in which 'Meteor' means lower atmosphere or a thing above land surface and "Logos" means to speak / discourse/ study.	This is also derived from the Greek word in which 'Klima' means slope of the earth (latitude) and '"Logos" means speak / discourse / study.
2.	A combination of both physics and geography.	A combination of meteorology and statistics.
3.	This science utilises the principles of physics to study the behaviour of air.	Climatology broadens the findings of meteorology in space and time.
4.	This science is concerned with the analysis of individual weather elements for a shorter period over a smaller area.	This science discusses the average conditions of weather over a longer period and larger area and their distribution.

From the foregoing information on the inter-relationships between meteorology and climatology it is clear that these sciences inevitably overlap and on many an occasion no sharp line can be drawn between them.

Agricultural Meteorology

Agriculture is defined as :

1. "The art and science of production and processing of plant and animal life for the use of human beings".
2. "A system for harvesting or exploiting the solar radiation".

Agriculture deals with three most complex entities viz., soil, plant and atmosphere and their interactions. Among these three, atmosphere is the most complex entity over the other two.

Agricultural meteorology is defined as :

1. "The study of those aspects of meteorology that have direct relevance to agriculture".
2. "A branch of applied meteorology which investigates the responses of crops to the physical conditions of the environment".
3. "An applied science which deals with the relationship between weather/climatic conditions and agricultural production".
4. "A science concerned with the application of meteorology to the measurement and analysis of the physical environment in agricultural systems".

The word 'Agrometeorology' is the abbreviated form of agricultural meteorology.

Agricultural Climatology

It is defined as, "A branch of applied climatology which studies the influence of climate on different processes of agriculture. "Agroclimatology" is the abbreviated form of agricultural climatology.

Practical Utility / Importance / Economic Benefits / Significance of Study of Agricultural Meteorology

In a broad manner the study of agricultural meteorology helps in :

1. Planning cropping systems / patterns.
2. Selection of sowing dates for optimum crop yields.
3. Cost effective ploughing, harrowing, weeding, etc.
4. Reducing losses of applied chemicals and fertilizers.
5. Judicious irrigation to crops.
6. Efficient harvesting of all crops.
7. Reducing or eliminating outbreak of pests and diseases.
8. Efficient management of soils which are formed out of weather action.

9. Managing weather abnormalities like cyclones, heavy rainfall, floods, drought, etc. This can be achieved by :
   (a) *Protection :* When rain is forecast avoid irrigation. But, when frost is forecast apply irrigation.
   (b) *Avoidance :* Avoid fertilizer and chemical sprays when rain is forecast.
   (c) *Mitigation :* Use shelter belts against cold and heat waves.
10. Effective environmental protection.
11. Avoiding or minimising losses due to forest fires.

Scope of Agricultural Meteorology

In addition to the points mentioned above, the influence of weather on agriculture can be on a wide range of scales in space and time. This is reflected in the scope of agricultural meteorology.

1. At the smallest scale, the subject involves the study of microscale processes taking place within the layers of air adjacent to leaves of crops, soil surfaces, etc. The agrometeorologists have to study the structure of leaf canopies which effects the capture of light and how the atmospheric carbon-dioxide may be used to determine rates of crop growth.
2. On a broader scale, agrometeorologists have to use the standard weather records to analyse and predict responses of plants.
3. Although the subject implies a primary concern with atmospheric processes the agrometeorologist is also interested in the soil environment because of the large influence which the weather can have on soil temperature and on the availability of water and nutrients to plant roots.
4. The agrometeorologist also be concerned with the study of glass houses and other protected environments designed for improving agricultural production.

ESTABLISHMENT OF A STANDARD METEOROLOGICAL OBSERVATORY

The India Meteorological Department (I.M.D) was established in the year 1875. At the time of starting, there were only 66 meteorological observatories or surface observatories spread throughout the country. The present number of observatories is more than 560. These observatories are meant for collection of data on various weather parameters such as temperature, humidity, wind, solar radiation, cloud formation, precipitation, etc.

To facilitate the collection of data at one point, five regional centres were established, which are located in five different places, for which the head quarters are mentioned against each as detailed below :

North zone	-	Delhi
East Zone	-	Kolkata
South Zone	-	Chennai
West Zone	-	Mumbai
Central Zone	-	Nagpur

The classification of the observatories is made based on the :

(a) Physical facilities available.

(b) Mode of transmission of data.

(c) The type of persons employed in the observatory.

There are six categories of observatories (Table 1.4).

In 1932 it was realised to collect the information on the influence of the weather parameters on different crops. So, most of the class I observatories were converted into agrometeorological observatories. These observatories are supervised by agencies like agricultural universities and research stations. There are 219 agromet observatories throughout the country.

Hours of Observation

As meteorological elements vary with time, it is necessary that they should be recorded at a particular time on every occasion.

In India, the main observations are recorded as per the guide lines of I.M.D at 0830 and 1730 IST.

At agromet observatories the observations are recorded at 0700 and 1400 hours LMT except evaporation and rainfall which are recorded at 0830 hours IST.

TABLE 1.4

CATEGORIES OF OBSERVATORIES

S. No.	Character	Category / Class					
		I	II	III	IV	V	VI
1.	Other name	Special station	Synoptic station	Synoptic station	Climatic station	Rainfall station	Non-instrument station
2.	Type of persons	Highly qualified and fulltime	Part time skilled workers	Part time skilled workers	Trained workers	Trained part time workers	Semiskilled workers
3.	Mode of Transmission	At once (on dot)	Two times a day	Once a day	Once a week	Once a month	On request
4.	Instruments	Automatic and sophisticated	Ordinary	Ordinary	Partially equipped	Rain-gauge	Nil

Radiation observations including the sunshine are recorded as per LAT (Local Apparent Time). At meteorological observatories the hours are numbered consecutively from midnight 00 hours to midnight 24 hours, the hours after noon being 13 hours, 14 hours and so on. Time as 2:30 p.m and 2:30 a.m. are expressed as 14.30 and 02.30 hours IST, respectively.

The instruments should be read in the following order in the observatory.

1. Wind instruments.
2. Raingauge.
3. Thermometers.
4. Barometer.

Non-instrumental observations i.e., clouds and visibility should be taken in 3 minutes intervals between first and second readings of the anemometer.

Selection of Site

The India Meteorological Department which is the principal government agency in all matters pertaining to meteorology, in India, recommends the following procedure in the selection of site for meteorological observatory.

1. The site should be well exposed, bare, levelled plot taking care of the proposed buildings, roads, canals, etc., which effect the exposure.
2. When the observatory is to be located in an agricultural farm, the site should be as far as possible so choosen that it is the representative of principal agricultural soils of the area.
3. In wooded areas, the site should be on an open plateau or terrace.
4. In areas of drifting sand the site should be at a place with minimum drift.
5. If the observatory is to be located near a reservoir or a lake, it must be located at the upwind of the reservoir, along the most prevalent direction of the high winds and at a distance away from drifts of spray from the spillways.
6. The site should be free from water logging, high structures, tall trees and hills, especially on the east and west as these cast shadows on the instruments. The above obstructions should be away by atleast ten times their height. Where radiation or sunshine observations are envisaged, there should be no substantial objects to the east or west subtending an elevation angle of more than 3 degrees with respect to the level of the radiation instrument or sunshine recorder.

7. The highest ground water level at the site must be less than two metres from the surface.
8. Final selection should be based on considerations such as proximity to a post office, water and the convenience to the observer.
9. The size of the observatory has to be decided depending upon the number of instruments to be installed. The general recommendation for size of the plot is :
   (a) 54 m NS by 36 m EW for an agro-meteorological observatory (Figure 1.3).
   (b) 25 m NS by 15 m EW for other purposes.
10. A suitable fencing should be provided.

Automatic Weather System (AWS)

A system capable of recording meteorological data at regular /defined interval, strong in memory and / or transmitting the same to a local or a remote location is called as automatic weather station or system.

Calculation of Time

A series of crossing lines on a map or a globe (Figure 1.4) which enable to identify the location of any point on the earth is known as the "Earth grid". A latitude of a place is the distance north or south of the equator which is measured as an angle whose apex is at the centre of the earth. One degree of latitude is approximately equal to 111 kilometres. The distance of a place east or west of the meridian of Greenwich or the Prime Meridian as an angle is known as longitude of a place. A meridian is a line joining places which have their noon at the same time.

Lines of latitude

These are the imaginary lines, circular in shape drawn horizontally connecting east and west on the globe and expressed in degrees. These are useful in calculation of distance of a place from the equator. The equator is imagined as zero and poles as 90° for this purpose.

Lines of longitude

These are also the imaginary lines connecting north and south on the globe and expressed in degrees. These are helpful in calculation of time.

Altitude

This is the height/angular height above the mean sea level.

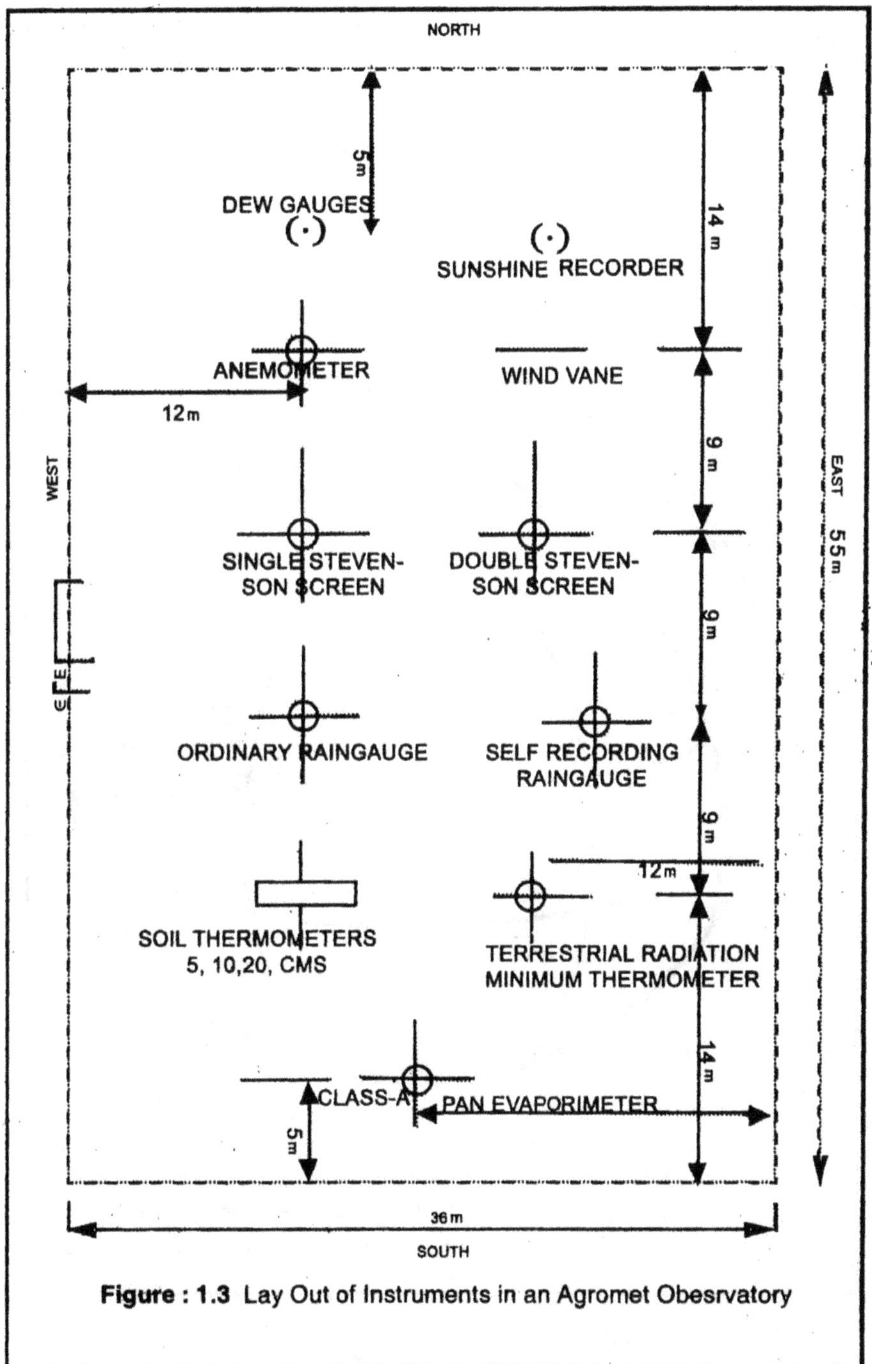

Figure : 1.3 Lay Out of Instruments in an Agromet Obesrvatory

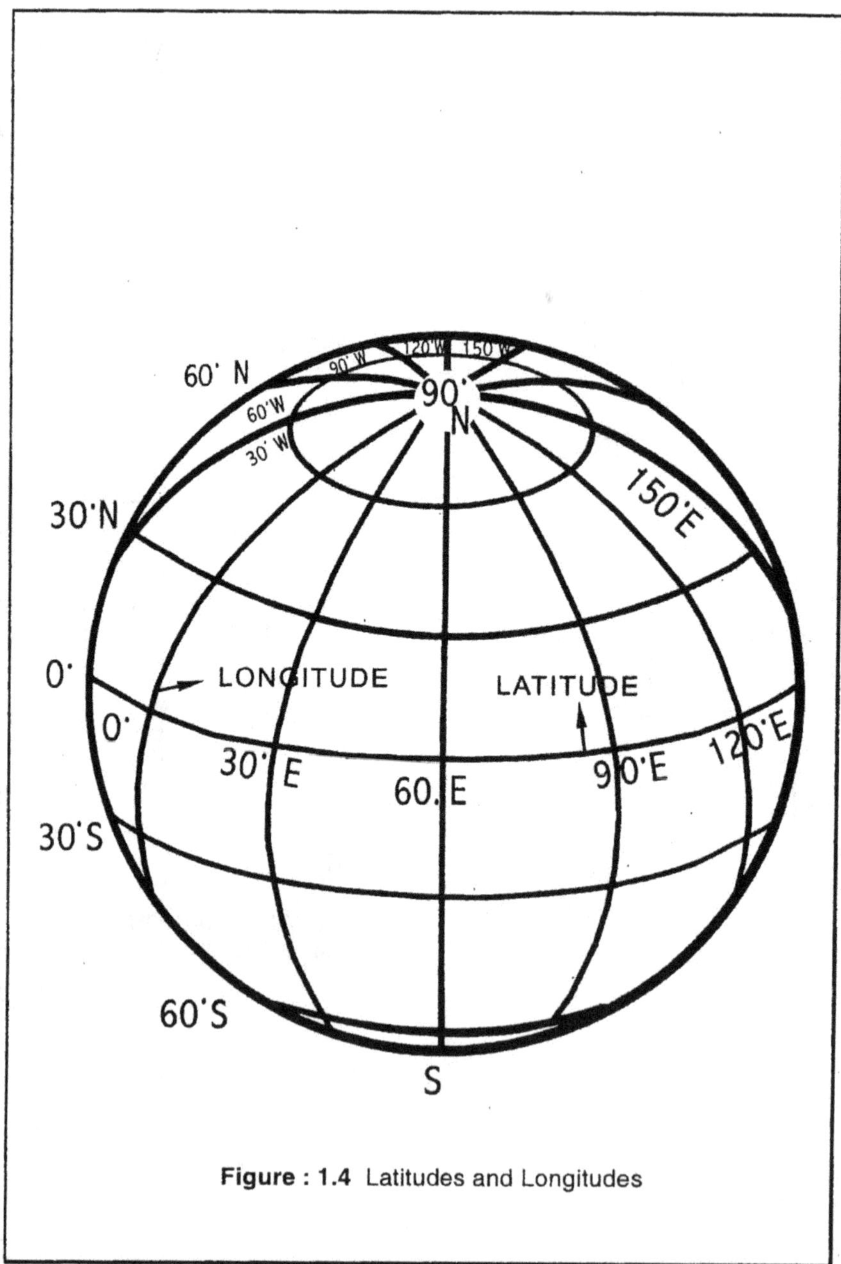

Figure : 1.4 Latitudes and Longitudes

TABLE 1.5

DIFFERENCES BETWEEN LONGITUDES AND LATITUDES

S. No.	Longitude	Latitude
1.	These are the lines running vertically to the equator connecting north and south.	These are the lines running horizontally to equator by connecting east and west.
2.	Prime meridian is taken as centre to divide the lines to east and west.	Equator is taken as centre to divide the other lines located in north and south.
3.	Useful in calculation of local mean time.	Useful for climatic study.
4.	Same line vary at different places on globe in numerical values.	Numeral values remain same on same line.

Indian Standard Time (IST)

The surface of the earth is divided into 24 "Time zones" the way in which there are 24 hours in a day. The time established in each of the zone is called as "Standard time". The Indian Standard Time (IST) is the Local Mean Time (LMT) for the meridian of longitude 82°31'E. This is the longitude of Allahabad which is taken as standard longitude for India. Since, each degree is equal to four minutes of time, IST is 5 1/2 hours ahead of Greenwich Mean Time (GMT). The GMT is also known as universal time.

Local Apparent Time (LAT)

The interval between two successive transits of the sun across the meridian is called the true solar day and the time based on the length of this day is called the Apparent Solar Time. Local Apparent Time is the apparent solar time for any particular place such that the sun passes across the geographical meridian at noon.

Local Mean Time (LMT)

This is the local time based on the transit of the mean sun. To calculate LMT from IST, it is essential to know the longitude of the station.

The Relation Between LAT, LMT and IST

$$LMT = IST- 4(L_1 - L_2)$$

Where L_1 = Standard meridian (82°30' for our country)

L_2 = The meridian of the station.

If the station is to the west of the standard meridian subtract 4 minutes for every degree from the IST. If the station is to the east of standard meridian, time has to be added to the IST.

$$LAT = LMT -- \text{Equation of time}$$

Values of correction due to equation of time which vary with season can be obtained from the tables (Since these are available in all the observatories, the student is advised to refer to them separately).

AGRICULTURAL SEASONS

Solstices and Equinoxes

During the revolution of the earth at certain time during the year it reaches maximum declination. So, the distance between the sun and the earth becomes maximum. Accordingly, the day and night are either longest or shortest . This is known as solstice (Figure 1.5). The winter solstice occurs on 22 December (Day is shortest and night longest) and the summer solstice on 21 June (Day is longest and night shortest).

Equinox is a moment at which the length of the day and night are equal in an year. The spring equinox falls on 21 March and the autumn equinox falls on 22 September (These dates of solstices and equinoxes are applicable only to northern hemisphere).

The changes in length of day and night causes changes in behaviour of weather and climatic elements. These changes divide the year into distinctive periods which are known as seasons.

The earth makes one complete rotation around itself in one day and around the sun in one year (364.25 days). It is tipped at an angle of 23.5° to its orbit in revolution. So, the northern hemisphere (also known as land hemisphere, on which maximum agricultural activity takes place) receives more sunlight and heat when the axis is tipped towards the sun

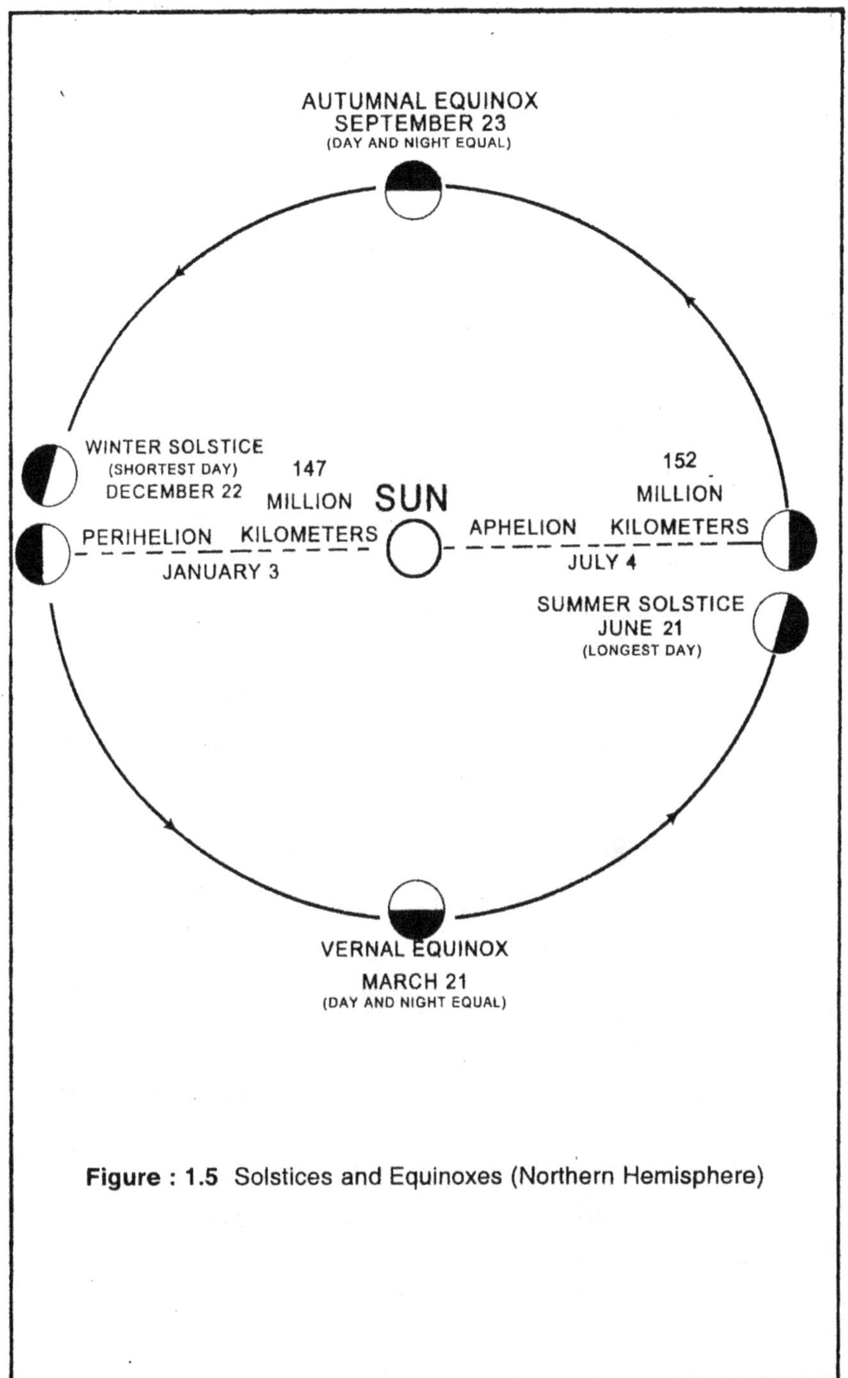

Figure : 1.5 Solstices and Equinoxes (Northern Hemisphere)

and less when the axis is tipped away from it. This movement of earth around the sun cause changes of seasons and crops are grown accordingly. In the northern hemisphere (temperate zone) the seasons are as follows;

1. Spring : March, April and May.
2. Summer : June, July and August.
3. Autumn : September, October and November.
4. Winter : December, January and February.

The India Meteorological Department divided an year into four seasons viz.,

1. Summer : March, April and May.
2. Monsoon : June, July, August and September.
3. Post-Monsoon : October and November.
4. Winter : December, January and February.

For agricultural purposes the year is divided into three crop seasons which are as follows :

1. Kharif or rainy season : From 15 June to 15 October.
2. Rabi or winter season : From 16 October to 15 February.
3. Summer season : From 16 February to 14 June.

The Agricultural Seasons of Andhra Pradesh

1. Kharif season (S-W Monsoon season) : June - September.
2. Rabi season (N-E Monsoon season) : October - February.
3. Summer season : February - May.

I. Kharif Season

1. This season is known as grand season / grand period of monsoon.
2. 70-80 per cent of the rainfall is received.
3. Highest rainfall is received in July and August.
4. Average temperature is 28°C.
5. Humidity is 60-80 per cent.
6. The major crops grown are castor, groundnut, cotton, maize, etc.

II. Rabi Season

1. This season is also known as retreating monsoon season.
2. 20 per cent rainfall is received.
3. The N-E monsoon enters from the Himalays to Chittor, and Nellore districts. So, these districts receive 40 per cent of total rainfall.
4. Due to cyclones and also due to N-E monsoon in October highest amount of rainfall is received.
5. Average temperature is 22°C.
6. Humidity is 30-50 per cent.
7. The major crops grown are wheat, safflower, etc.

III. Summer Season

1. No rains are received. Once in a while rains due to convection may occur (now-a-days cowpea crop is grown)
2. Temperature is 40°C.
3. Humidity is 10 - 20 percent.

Geographical Regions of Andhra Pradesh

There are three geographic regions in Andhra Pradesh. In each region the seasons are named in different ways.

1. Coastal Region

A. Punasa : April, May - August, September.

B. Pyru : November - March.

2. Telangana Region

A. Kharif/Early (Abi) : June - October.

B. Rabi/late (Tabi) : November - April.

3. Rayalaseema Region

A. Mungari/Early : June - October.

B. Hingari/Late : September - March.

Peddapanta is the word used for late kharif in Srikakulam, Vizianagaram and Vishakapatnam districts. Sarwa and Dalwa are the seasons respectively for early and late season crops with reference to rice in the Coastal districts.

Chapter - 2

Solar Radiation

"O my Lord, O primeval philosopher, maintainer of the universe, O regulating principle, destination of the pure devotees, well-wisher of the progenitors of mankind, please remove the effulgence of your transcendental rays so that I can see Your form of bliss. You are the eternal Supreme Personality of God-head, like unto the sun, as am I".

Solar radiation is the primary source of energy on earth, and life depends on it. Solar radiation is defined as, "The flux of radiant energy from the sun". All matter at a temperature above the absolute zero, imparts energy to the surrounding space. This energy is transformed by green plants in the process of photosynthesis into the potential energy of organic material. In inorganic bodies the rays absorbed are used in heating. The variations of the total radiation flux from one site to another on the surface of the earth are enormous and the distribution of plants and animals responds to this variation.

Sun

The main features of the sun are :

1. The sun forms nucleus of the solar system.
2. All planets revolve around the sun in elliptical orbits.
3. It looks bright and big because of its nearness to the earth than the other stars.
4. It is at a mean distance of 150 million kilometres from the earth.
5. Its diameter is 13,92,000 kilometres (the diameter of earth is 12,756 kilometres).
6. Its volume is approximately 1.3 million times more than that of the earth.
7. Eventhough it is gaseous, it weighs more than 300,000 times as much as the earth.
8. It takes 8 minutes 20 seconds for its rays to reach the earth.
9. The solar surface is composed of three gaseous layers :
   (a) The inner most surface layer is called the "Photosphere" and its temperature is 6000°C.
   (b) The middle surface layer is called the "Chromosphere" and its temperature is 5000°C.
   (c) The outer surface layer is called the "Corona" and its temperature slightly varies.

Insolation

The abbreviated form of **in**coming **sol**ar radi**ation** is "Insolation". It consists of bands of radiant energy of different wavelengths.

Processes of Transmission of Energy

Heat energy is transmitted by three processes.

1. Radiation

This is the process of transmission of energy from one body to another without the aid of a material medium (solid, liquid, gas, etc.).

Example : The energy transmission through space from the sun to the earth.

2. Conduction

This is the process of heat transfer through matter without the actual movement of molecules of the substances or matter. Heat flows from the warmer to cooler part of the body so that the temperature between them are equalised.

Example : The energy transmission through an iron rod which is made warmer at one end.

3. Convection

This is the process of transmission of heat through actual movement of molecules of the medium. This is the predominant form of transmission of energy on the earth as all the weather related processes involve this process.

Example : Boiling of water in a beaker.

Importance of Solar Radiation on Crop Plants

1. From germination of seed to harvesting and even post harvest processes are affected by solar radiation.
2. Solar radiation provides energy for :
   - All the phenomena related to biomass production.
   - All photosynthetic processes.
   - All physical processes taking place in the soil, the plant and their environment.
3. Solar radiation controls the distribution of temperature thereby distribution of crops into different regions.

Basic Radiation Laws

Wavelength

The wavelength of electromagnetic radiation is given by the equation :

$$\lambda = \frac{C}{\nu}$$ where,

λ = Wavelength (The shortest distance between consecutive crests in the wave trance)

C = Velocity of light (3×10^{10} cm Sec^{-1})

ν = Frequency

Planck's Law

The electromagnetic radiation energy content (E) of each quantum is proportional to the frequency given by the equation :

$E = h\nu$ where,

h = Planck's constant (6.625×10^{-27} ergs Sec^{-1})

ν = Frequency

The law states that greater the frequency greater is the energy of quantum.

Kirchoff's Law

A good absorber of radiation is a good emitter, in similar circumstances. This law states that the absorptivity 'a' of an object for radiation of a specific wavelength is equal to its emissivity 'e' for the same wavelength. The equation of the law is :

$a(\lambda) = e(\lambda)$

Stefan Boltzmann's Law

The intensity of radiation emitted by a radiating body is directly proportional to the fourth power of the absolute temperature of that body.

Flux = σT^4 where,

T = (273 + °C) because temperature is in Kelvins

σ = Stefan Boltzman's constant, which is equal to 5.67×10^{-5} ergs cm^{-2} Sec^{-1} K^{-4}

Wein's Displacement Law

The wavelength of the maximum intensity of emission (λ max) from a radiating black body is inveresly proportional to its absolute temperature (T).

λ max = 2897 T^{-1}

If the temperature of a body is high, radiation maximum is displaced towards shorter wavelengths. For the surface temperature of 5793°K, evaluated for the sun λ_m is 0.5. The most intense solar radiation occurs in the blue-green range of visible light. The wavelength of maximum intensity of radiation for the earth's actual surface temperature of 14°C or 287°K is about 10.0 microns, which is into infrared band.

Solar Spectrum

Radiant energy is transmitted in the form of electromagnetic waves by the sun. The energy from the sun is spread over a very broad band of wavelengthes known as solar spectrum. It is also known as electromagnetic spectrum. The spectrum does not constitute only one band, but a combination of different waves which are characterised individually (Figure 2.1).

Example : U.V. rays, light part, Near I.R., Far I.R., Radio waves, micro waves, radar waves, etc.

TABLE 2.1

ENERGY CONTENT OF DIFFERENT BANDS IN SOLAR SPECTRUM

Band No.	Spectrum	Wavelength in Microns	% of Energy
1.	γ - rays & X - rays	0.005 - 0.20	9 %
	U. V. rays	0.20 - 0.40	
2.	Violet	0.40 - 0.43	41 %
	Blue	0.43 - 0.49	
	Green	0.49 - 0.53	
	Yellow	0.53 - 0.58	
	Orange	0.58 - 0.63	
	Red	0.63 - 0.70	
3.	Infrared rays	>0.70	50 %

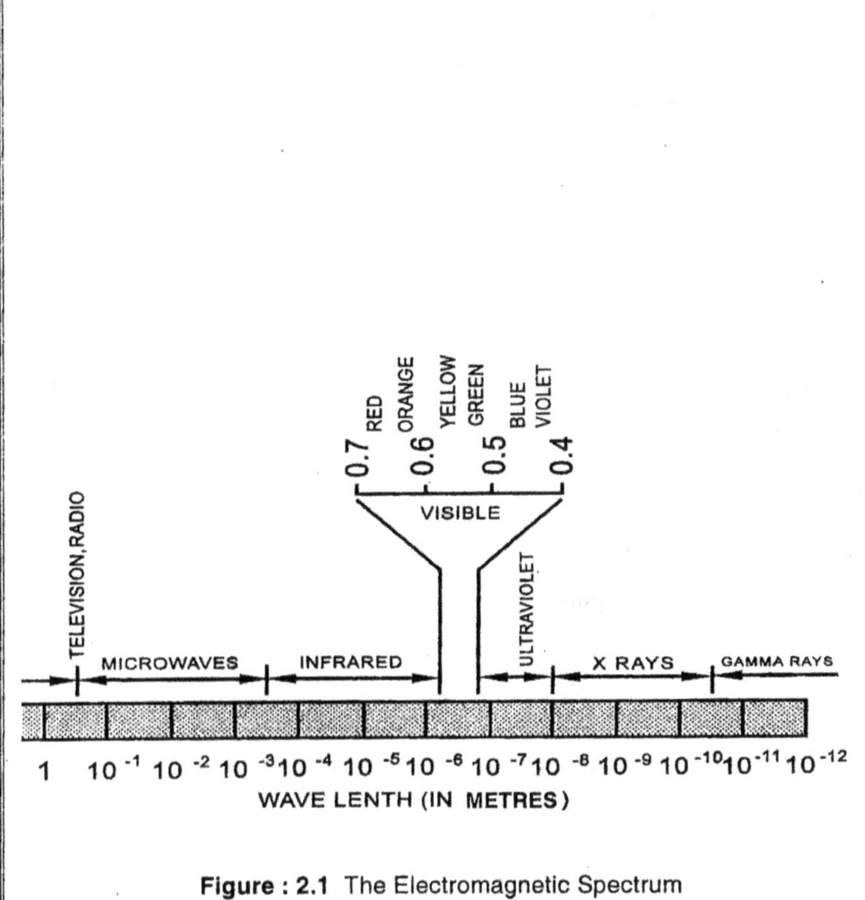

Figure : 2.1 The Electromagnetic Spectrum

Different Bands of Solar Spectrum

1. The shorter wavelengths of the spectrum are known as ultra - violet rays (U. V rays). These are chemically very active. Unless, these are filtered in the atmosphere, there is a danger for life on the earth. This band ranges between 0.005 to 0.4 microns.
2. The part of the spectrum which is visible is known as 'Light'. It is the part of the spectrum which is essential for all the plant processes and ranges from 0.4 to 0.7 microns.
3. The third part of the solar spectrum (last band) is known as infrared band. This is essential for thermal energy of the plant (the source of heat to the plant). This band is > 0.7 microns.

Functions of Light

The functions of light are :

1. All the plant parts are directly or indirectly influenced by light.
2. Light of correct intensity, quality and duration is essential to normal plant development.
3. Poor light availability causes abnormalities and disorders in plants.
4. Light is indispensable to photosynthesis.
5. Light governs the distribution of photosynthates among different organs of plants.
6. Effects tiller production.
7. Effects stability, strength and length of culms.
8. Effects dry matter production.
9. Effects the size of the leaves.
10. Effects the root development.
11. Effects the flowering and fruiting.
12. Effects the dormancy of the seed.

Characteristics of Light

The most important three characteristics of light which effect the plant growth and development are :

I. Intensity II. Duration III. Quality (wavelength)

I. Light Intensity

1. Light is the energy source for the photosynthetic process. The human eye can detect solar radiation in the wave band 0.4 to 0.7 microns. This process is called as vision, and solar radiation in this wave band, when concerned with vision is called the "Light". Eventhough, photosynthesis is driven by solar radiation from 0.35 to 0.95 microns the "Light" is the energy source for photosynthetic processes.
2. There is a linear relationship between the light intensities and rate of photosynthesis at low light intensities.
3. With increasing light intensity, photosynthesis of a single leaf decreases.
4. Extremely high light intensities may have an inhibitory effect on photosynthesis, a phenomena called polarization or photo-oxidation in which cell contents are oxidised by atmospheric oxygen.
5. Under field conditions the light is not spread evenly over the crop canopy but commonly passed by reflection and transmission through several layers of leaves.
6. The intensity of light falls at exponential rate with the path length through absorbing layers according to Beer's law. This law states that the relative radiation intensity decreases exponentially with increasing leaf area.
7. The intensity of light falling on the leaves in the lower layers depends on several factors.
8. At ground level the final light intensity is usually below compensation point. Light compensation point is defined as, "The light intensity at which the gas exchange resulting from photosynthesis is equal to that resulting from respiration".

II. Duration of Light

1. The duration of light has a great influence on the process of crop canopy development and final yield.
2. Some crops react differentially to durations of day and night. In some cases the fruiting is hastened and in other cases it is delayed.

3. The influence of relative length of day and night on plants is called the "Photoperiodicity".

4. Based on the photoperiodic requirements for floral initiation the crop plants are divided into :

   (a) *Long day plants :* Plants which develop rapidly and flower early when days are long (more than 10 hours)

   Example : Crops grown in higher latitudes like wheat, rye, oats, barley, flax, clover, etc.

   (b) *Short day plants :* Plants which develop rapidly and flower early when days are short (nights are more than 10 hours).

   Example : Crops grown in tropics like cotton, corn, beans, sunflower, tomato, cucumbers, etc.

   (c) *Day neutral plants :* Plants which require 12-14 hours of light daily to initiate flowering.

   Example : Most of the perennials.

5. Recent research revealed that the duration of night is often more important than the length of day light. Photoperiod changes in a regular manner with latitude and season.

6. Most parts of India receive long hours of sunshine as also congenial temperatures. If sufficient soil moisture is made available, then optimum crop growth and highest final seed yields of crops can be obtained.

7. In agro-meteorolgical research a 24 hour combination of light and dark usually constitutes one cycle. A short day cycle often consists of 8-10 hours of light and 14-16 hours of dark and vice-versa for a long day cycle.

8. A short day plant when given light for a short period in night it will not flower because the response of short day plants is more a response to the dark period than to "Light". But, a long day plant subjected to similar situation will flower.

9. The varying photoperiodicity of plants is a very important factor in crop rotation which helps not only in operational nutrient management but also in disease eradication.

10. Choice of varieties and cultivars in newly cultivated areas depend on duration of light.

III. Quality of Light

1. The quality of solar radiation affects several events in crop plant life. The red part of light (0.66 microns) play an important role. It :
   (a) is the most effective inhibitor of flowering in long day plants,
   (b) helps mature apples to turn red,
   (c) induces germination,
   (d) supresses elongation of stem, and
   (e) has weakest influence on phototropism.
2. The blue part of light has strongest influence on phototropism. The blue and green part of light inhibits germination of seeds.
3. Farred part of the spectrum promotes elongation of stem.
4. The U.V. part of the spectrum is harmful to the crop plants (Germicidal effect). However, it has some beneficial effects like killing micro-organisms, disinfecting the soil, disease eradication, etc.
5. Between 0.75 and 2.1 microns the leaf absorbs weakly, thereby escaping the radiation load from solar radiation. The energy in this wave band is not effective in photosynthesis. The main effect is thermal and respiration is encouraged.

Reflection, Transmission and Absorption

On an average about 75 per cent of the incident radiation is absorbed by the plant canopy. About 15 per cent is reflected and 10 per cent is transmitted. Reflection and transmission from the leaves have similar spectral distributions (Figure 2.2).

1. The maxima for both reflection and transmission is in the green light as well as in the infrared region.
2. The strong infrared reflection from plants is an important natural device for protection of plant life against damage due to overheating.
3. Due to their chemical components or physical structures, plants absorb selectively in discrete wavelengths.
4. Chlorophyll absorption is maximum in the blue (0.45 microns) and in the red regions (0.65 microns).

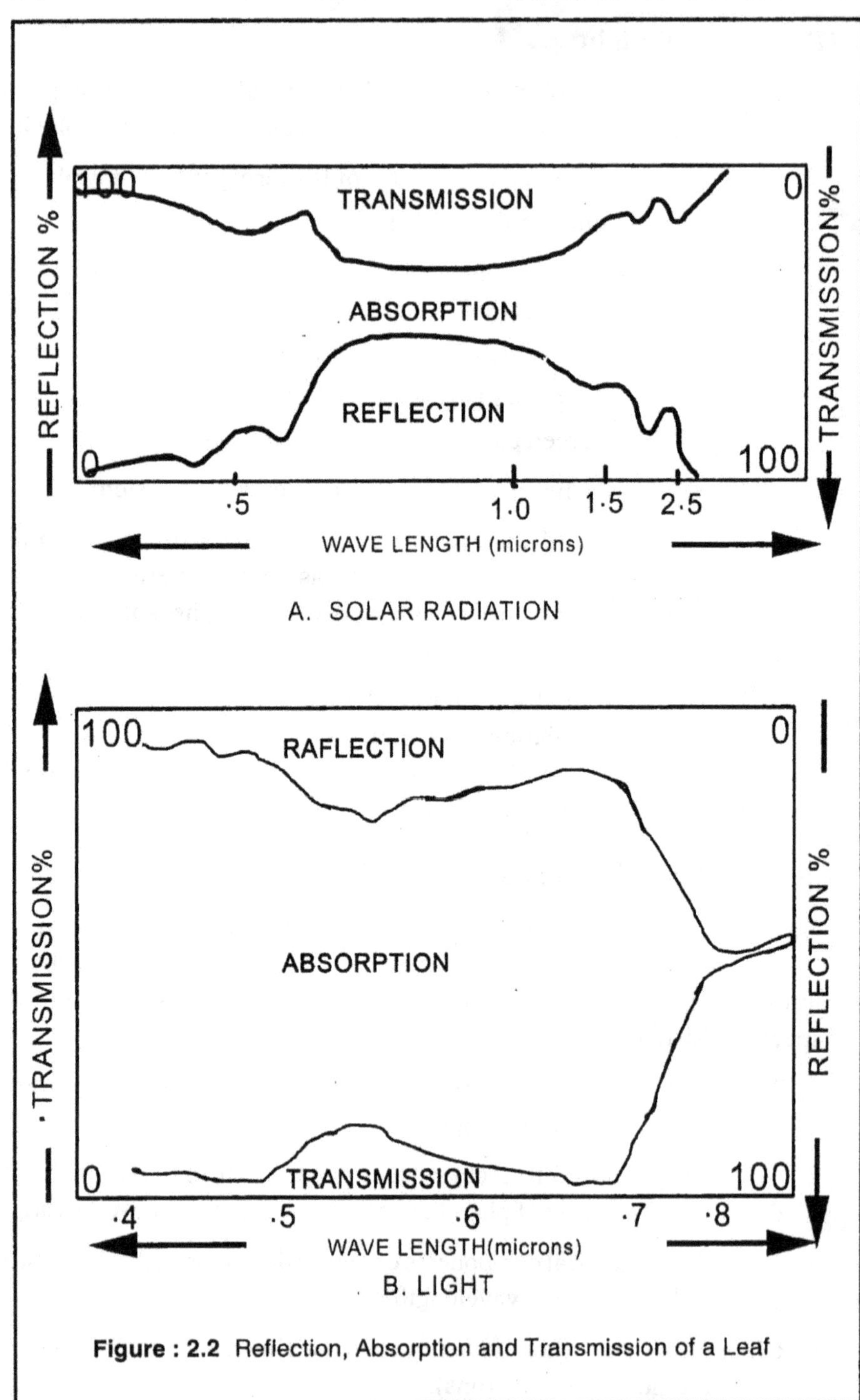

Figure : 2.2 Reflection, Absorption and Transmission of a Leaf

5. Near the border of visible light, absorption by the plants decreases but again increases in the infrared.

6. Infrared radiation greater than 3 microns is completely absorbed by the plants.

Factors Affecting the Solar Radiation Receipt on the Earth Surface

The amount of solar energy received on the earth is affected by the following factors.

I. Astronomical Factors

1. *Solar output* : The sun behaves like a black body. So, the energy radiated by the sun depends on its temperature in accordance with the Stefan's law. Assuming a temperature of 5800°K for the sun the solar output to space is 2.33×10^{25} $kJmin^{-1}$. But, only a tiny fraction of this is intercepted by the earth. The reason is that the energy received is inversely proportional to the square of the solar distance i.e., 150 million kilometres.
2. *Distance between the earth and the sun* : Due to eccentricity of the earth's orbit around the sun, the distance of the earth from the sun changes. Consequently, the receipt of energy on the earth surface changes which brings seasonal changes. The receipt of solar energy on a surface normal to the beam is 7 per cent more on 3 January at the perihelian than on 4 July at the aphelion.
3. *Altitude of the sun :* The vertical rays of the sun heats the minimum possible area. The greater the sun's altitude the more concentrated is the radiation intensity per unit area at the earth's surface.

II. Geographical Factors

1. *Effect of atmosphere* : The quantity of solar radiation passing to the earth is depleted due to the presence of aerosols. Similarly, clouds also scatter, reflect and absorb solar radiation. So, the amount of radiations reaching the earth is affected.
2. *Effect of latitude* : According to latitude the amount of solar energy reaching the earth changes because the day length and the distance travelled by inclined sun rays through the atmosphere changes.

3. *Effect of aspect and altitude* : The flux of solar radiations received increase with increasing altitude. Mountain areas which faces sun rays receives more radiation while the valley floor may receive less amount.

4. *Length of the day* : As the duration of the day increases the amount of radiation also increases. The intensity and the day duration together decide the quantity of radiation receipt.

Factors Affecting the Distribution of Solar Radiation within the Plant Canopy

1. Type of Plants

(a) The leaves of cereal crops like paddy, wheat, etc., have a transmissivity from 5 to 10 per cent.

(b) The broad leaves of ever green plants have lower value of 2 to 8 per cent, whereas aquatic plants have 4 to 8 per cent.

(c) Transmissivity changes with the age of a leaf.

2. Age of the Leaves

The transmissivity of younger leaves is more as compared to older leaves.

3. Chlorophyll Content

As the chlorophyll content increases the values of transmissivity decreases.

4. Arrangement of Leaves

(a) The relative light interception by horizontal and erect foliage is 1 : 0.44.

(b) When the leaf area index is one (1) the light transmissivity of more upright leaves is 74 per cent as against 50 per cent for horizontal leaves.

5. Angle of Leaves

(a) In full sunlight, the optimum inclination for efficient light use is 81°.

(b) At full sunlight, a leaf placed at the optimum inclination is 4-5 times as efficient in using light as a horizontal leaf.

(c) The ideal arrangement of leaves (shall be) is that the lowest 13 per cent of the leaves lay at angles between 0° and 30° to the horizontal, that the adjoining 37 per cent of the leaves lay at 30° to 60° and the upper 50 per cent of the leaves lay at 60° to 90°.

6. Plant Density

In case of sparse crop stands not only the per cent of light transmissivity is more but it is also variable with the time of the day. It is minimum at noon and maximum during morning and evening hours. In dense crop canopies the light transmissivity is less.

7. Plant Height

When the plant height increases the transmissivity of light by the canopy decreases.

8. Angle of the Sun

(a) The highest radiation penetration occurs at noon.

(b) Relatively high radiation penetration also occurs both in the morning as well as before the sunset due to high proportion of diffuse light.

Physiological Responses of Plants to Different Bands of Incident Radiation

The Dutch Committee on plant irradiation has divided the solar spectrum into eight bands. This was done on the basis of the physiological response of plants to the incident radiation (Table 2.2).

Measurement of Solar Radiation

Radiometry is the science of measurement of radiant energy. An instrument which measures radiant energy is called as radiometer. The instruments to measure various spectral bands of solar radiation should have the following qualities :

TABLE 2.2

PHYSIOLOGICAL RESPONSES OF PLANTS TO DIFFERENT BANDS OF INCIDENT RADIATION

S. No.	Band No.	Spectral Region (microns)	Character of Absorption	Physiological Effect
1.	1st	Infrared >1.000	By water in tissues	Converted into heat, this has no specific effects on photochemical and biochemical processes.
2.	2nd	1.000 to 0.700	slight	Stimulates elongation in plants.
3.	3rd	0.700 to 0.610	Very strong by chlorophylls	Large effect on photosynthesis and photoperiodism.
4.	4th	0.610 to 0.510	Somewhat less	Small effect on photosynthesis and small morphogenic effect.
5.	5th	0.510 to 0.400	Very strong by chlorophylls and carotenoids	Large effect on photosynthesis and large morphogenic effect.
6.	6th	0.400 to 0.315	By chlorophylls and protoplasm	Small effect on photosynthesis. Produces fluorescence in plants.
7.	7th	0.315 to 0.280	By protoplasm	Significant germicidal action, Large morphogenic effect and Large effect on physiological processes.
8.	8th	< .280	By protoplasm	Large germicidal effects. Lethal in large doses.

A. Normal

- Reliability.
- Reproducibility.
- Stability.
- Insensitivity to other environmental factors.

B. Specific

- Responsiveness to the wavelength being studied.
- Response of the sensor must be similar to that of the plant.
- Cosine response.

Instruments to measure components of radiation

I. Radiation Flux Measuring Instruments

The following instruments measure different components of solar radiation mentioned against each one of them.

1. *Pyranometer :* Total shortwave radiation or components on a horizontal plain surface from hemispherical sky.
2. *Net pyranometer or solarimeter :* Net shortwave radiation in W m^{-2}.
3. *Pyrradiometer :* Global or total radiation from sky on a horizontal surface in W m^{-2}.
4. *Net pyrradiometer or net radiometer :* Net shortwave and longwave radiation in W m^{-2}
5. *Pyrheliometer :* Direct solar beam on a plane surface at normal incidence in W m^{-2}
6. *Pyrgeometer :* Net infrared radiation of the atmosphere in W m^{-2}.
7. *Quantum sensor :* Number of photons received per unit area in E m^{-2} S^{-1}.

II. Measurement of Illumination or Brightness

The brightness is measured with luxmeter or photometric sensors in foot candles or lux. To measure light the instruments used are light meters (photometer).

III. Sunshine Duration or Cloudiness Hours

In most of the meteorological observatories, the duration of sunshine is recorded. These are measured with sunshine recorder (The units are hours). However, solar radiation is also recorded simultaneously in few of these observatories. There is a relationship between the solar radiation and the duration of sunshine.

$$Q/Q_A = (a+b)\ n/N$$

Where,

Q = The radiation actually received.

Q_A = Angot's value.

n = Actual duration of sunshine received.

N = Maximum possible duration of sunshine.

a and b = Constants i.e., 0.23 and 0.48 respectively.

Note : Angot's value denotes the theoretical amount of radiation that would reach the surface of the earth in the absence of atmosphere. The values of 'a' and 'b' vary according to longitude and latitude.

The number of hours of bright sunshine is recorded by a number of instruments. Of all, the Campbell - Stokes Sunshine Recorder is the best one.

Campbell - Stokes Sunshine Recorder

Principle :

Sun's rays are concentrated on a chemically sensitised card by a spherical lens. This card produces a trace as the sun rays fall during the hours of bright sunshine. As the card is graduated in hours and tenths, the daily duration of sunshine can be easily determined.

Operation and Measurement

1. This is an instrument for recording the duration of bright sunshine.
2. It consists of a glass sphere fixed centrally to a frame. Just below the glass sphere a hemispherical bowl is rigidly fixed to the frame. The bowl has three slots or grooves through which the chemically treated cards are inserted. The frame is mounted on a base provided with three levelling screws (Figure 2.3).

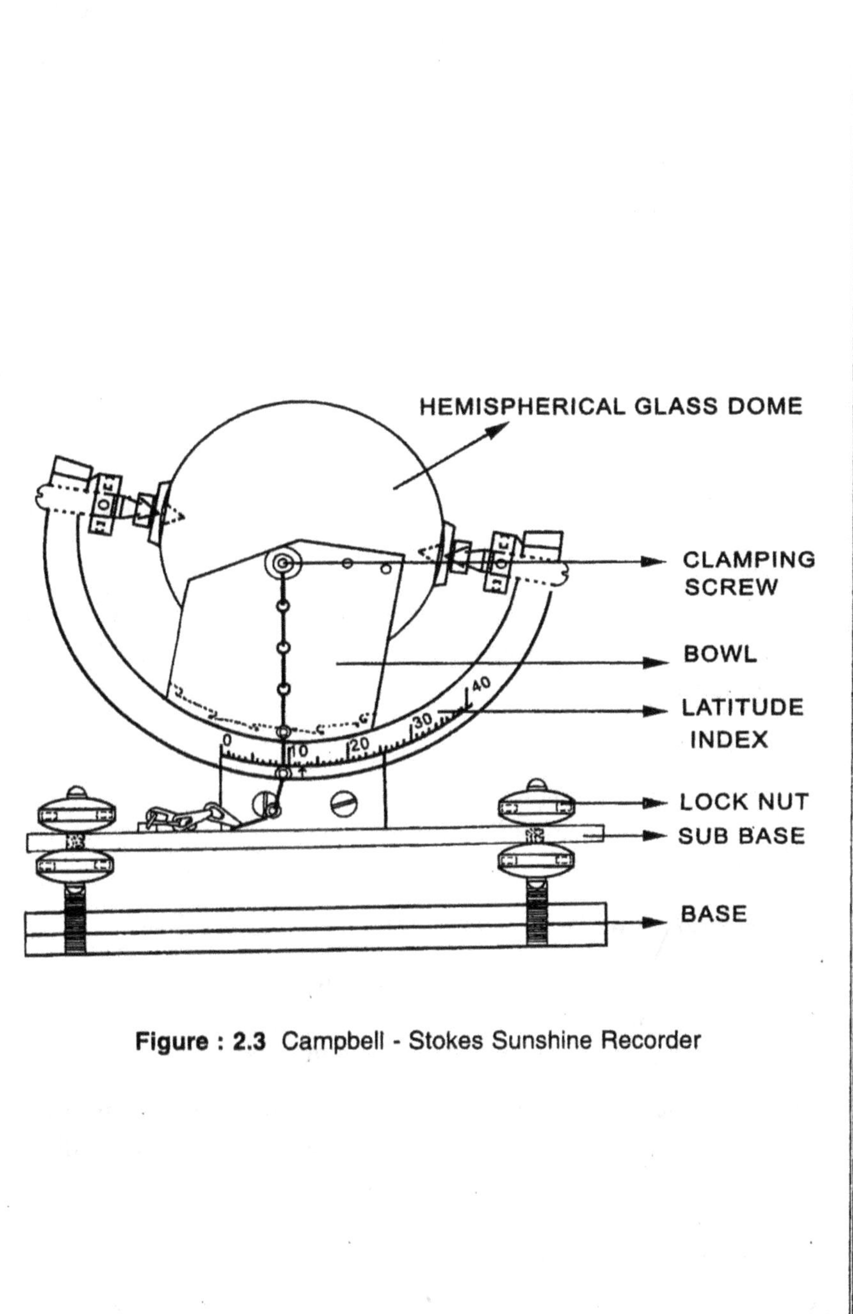

Figure : 2.3 Campbell - Stokes Sunshine Recorder

3. The glass sphere acts as a converging lens. The different points on chemically treated cards represent the principle foci for the different positions of the sun, during the apparent movement of earth, from east to west.
4. The bright sun rays leave a charred or burnt line on the chemically treated card. The cards are graduated in hours for accurate measurement of bright sunshine.
5. The sunshine recorder is kept on a platform at a height of 10 feet from the ground surface.
6. It is kept in a perfectly horizontal plane. To achieve this, the levelling screws are adjusted and if needed a spirit level can also be used to bring the instrument in a perfectly horizontal position.

There are three types of cards available for measuring bright sunshine.

(a) *Long Curved Cards*

These are also called as "Summer cards" and these are used from 13th April to 31st August. These cards are introduced through the bottom slot in the concave plate.

(b) *Short Curved Cards*

These are also called as "Winter cards" and are used from 13th October to the end of February. These are introduced at the top slot.

(c) *Straight Cards*

When day and night lengths are equal, these cards are used i.e., from 1st March to 12th April and 1st September to 12th October. These are introduced through the middle slot.

In recent models of this recorder, only one slot for all season cards has been implemented by having a broader card slot, thereby enabling different curvature cards for different season periods.

Basic Radiation Terminology

Shortwave Radiation

This is a part of the solar spectrum having wavelength less than 4 μ (microns).

Longwave Radiation

The terrestrial and atmospheric radiation which are in the wavelengths between 4 and 120 microns.

Net Radiation

The difference between the incoming radiation from the sun and the out going solar radiation from the earth is known as net radiation. The net radiation values become -ve after late evening hours to early morning hours. It is a conservative term and plays an important role in the energy processes of the crops.

Solar Constant

It is the energy falling in one minute on a surface of 1 cm^2 at the outer boundary layer of the atmosphere, held normal to the sunlight at the mean distance of the earth from the sun.The units are cal/cm^2/min. "Cal/cm^2 " is also known as "Langley". The estimated value of this constant varies from 1.94 to 2.0 Langely/min. The average value is 2 Langely/mn..

It depends on :

1. Output of solar radiation.
2. Distance between the earth and the sun.
3. Transparency of the atmosphere.
4. Duration of the sunlight period.
5. The angle at which the sun's rays strike the earth.

Black Body

It is an ideal hypothetical body which absorbs all the electromagnetic radiation falling on it. It neither reflects nor transmits any radiation striking it. However, when heated, it emits all the possible wavelengths of solar radiation and becomes a perfect radiator. So, an ideal black body is a perfect absorber and a perfect radiator.

Black Body Radiation

The radiation radiated by an ideal black body is known as black body radiation.

Emittance

It is the ratio of the emitted radiation of a given surface at a specified wavelength to the emittance of an ideal black body at the same wavelength and temperature. For other than a black body the value of emittance is always less than one and for black body the emittance value is one.

Absorptivity

For an object, this is the ratio of the electromagnetic radiant power absorbed to the total amount of radiation incident upon the same object. Like emissivity the values are less than one for other than a black body and one for a black body.

Reflectivity

The ratio of the monochromatic beam of electromagnetic radiation reflected by a body to that incident upon it. The units of expression is per cent.

Transmissivity

This is the ratio transmitted to the incident radiation on a surface preferably a crop canopy.

Albedo

It is the ratio between the reflected radiation to the incident radiation on a crop field, snow, leaves, etc. For white bodies the albedo values are high. For fresh snow cover the albedo values range between 75 and 95; for cropped fields it is 12-13; dark cultivated soil 7-10; human skin 15-25, etc. Albedo determines how much of heat that reaches the ground in the form of radiation will remain available for use.

Results of Research on Effect of Solar Radiation on Important Crops

I. Rice

Rice is the staple food for people in nearly all Asian countries and a major source of livelihood in their rural economies. World rice production, consumption and trade are concentrated in Asia. Asia contributes 90 - 95 per cent of world production. This crop occupies second place next to wheat. Among various countries that grow rice, in 3 countries the yields are >6 t/ha; 17 countries 4 t/ha; 78 countries <3 t/ha; 58 countries <2 t/ha and in 20 countries it is even <1 t/ha. This large variation in yield between <1 t/ha to > 6 t/ha is mostly due to environmental factors.

1. The photosynthetically active radiation (PAR) is in between 0.4 and 0.7 microns which is also referred to as light. The PAR available at the canopy, its interception and utilization by the crop determine to a large extent the yield and quality in rice.
2. The solar radiation requirement of rice crop differs from one growth stage to another. Shading at vegetative stage slightly affects the yield and yield components. In addition to this, stressed environments results in net decrease in crop dry matter.
3. But, shading at reproductive stage, however, has a pronounced effect on spikelet number; and during ripening reduces the yield considerably because of reduction in percentage of filled spikelets. An average of 300 cal/cm^2 during reproductive stage makes yields of 5 t/ha possible.
4. In the tropical regions the solar radiation is higher in the dry than in the wet season. Consequently, the dry season yields are higher.
5. The excessively cloudy weather during wet season is often considered a serious limiting factor for rice production in monsoon Asia.
6. Another major utility of the solar radiation is the solar drying of paddy seeds and husk. Eventhough, this practice is traditional it is more economical than forced convection driers.

II. Groundnut

1. Groundnut responds to full light intensities. The clear, cloudless days are advantageous for maximum photosynthesis and high yields can be obtained on such days.
2. Low light intensity prior to onset of flowering slows down vegetative growth. During rapid growth low light intensity increases height and length of stems, but decreases leaf weight and flowering.
3. Low light intensities suppresses development and growth of reproductive branches and consequently decrease the total flower production.

4. In early flowering stage low intensity of light causes abortion of flowers and at pegging stage significantly reduce peg number and pod weight.
5. Low light intensity during pod filling and maturity stages slightly decreases number and weight of mature pods but significantly increase percentage of shrivelled kernels.
6. Shading before onset of flowering slows vegetative growth. Shading in early flowering stage appeared to affect distribution of pods around mainstem. Shading after peak flowering interrupted in filling, resulting in a significant reduction in percentage of extra large kernels.
7. Solar radiation detoxify groundnut oil contaminated with aflatoxin.
8. Highly significant negative linear relationship exist between the night temperature and the radiation use efficiency.
9. The radiation use efficiency is negatively associated with canopy extinction coefficient.
10. Solarisation during hot summer will provide a sufficient level of suppression on root-knot nematodes.

III. Sugarcane

1. Abundant solar radiation is required for accumulation of sucrose.
2. Plants subjected to full sunshine are thicker, have dark green foliage and thus high sucrose as compared to non-exposed crop canopies.
3. When the solar radiation contains 50 per cent of PAR, the crop produces more number of millable canes, high sucrose content and increase in weight which is vice-versa under reverse conditions.

IV. Cotton

1. Sun radiation directly affects temperature, and the plants need an exceptional amount of sunlight; 50 per cent of normal radiation is considered to be the lowest limit for successful cultivation of cotton.
2. Regions recording 60 pre cent cloudiness are absolutely unsuited to successful growing. The sunshine is indispensable for vegetative development, formation and ripening of bolls, etc.

3. Cloudiness affects fruit setting, and cloudy weather lasting 2 to 8 days causes speedy shedding of fruit. The greatest assimilation activity naturally takes place in full sunlight during early morning, with stomata fully open.

Units of Measurement

TABLE 2.3

TERMS AND UNITS IN RADIATION MEASUREMENT

Term	Units
Radiant energy	J
Number of photons	mol
Radiant exposure	$J\ m^{-2}$
Photon exposure	$mol\ m^{-2}$
Radiant flux	$J\ s^{-1} = W$
Photon flux	$mol\ s^{-1}$
Radiant flux density	$W\ m^{-2}$
Photon flux density	$mol\ m^{-2}\ s^{-1}$
Irradiance	$W\ m^{-2}$
Incident photon flux density	$mol\ m^{-2}\ s^{-1}$
Emittance	$W\ m^{-2}$
Radiant intensity	$W\ sr^{-1}$

Chapter - 3

Temperature

"Although fixed in His abode, the Personality of God-head is swifter than the mind and can overcome all others running. The powerful demigods cannot approach Him. Although in one place, He controls those who supply the air and rain. He surpasses all in excellence".

AIR TEMPERATURE

Temperature is defined as, "The measure of speed per molecule of all the molecules of a body" whereas heat is, "The energy arising from random motion of all the molecules of a body".

The temperature of a body is the condition which determine its ability to transfer heat to other bodies or to receive heat from them. In a system of two bodies the one which looses heat to the other is said to be at a higher temperature.

Heat measures total molecular energy. Temperature measures average energy of individual molecules. Temperature is that characteristic of a body which determines the direction of heat flow by conduction.

Importance of Air Temperature on Crop Plants

1. Temperature influences distribution of crop plants and vegetation (In Western Himalayas the temperature falls as altitude increases and this change is responsible for the change of vegetation at different altitudes).
2. The growth and development of crop plants are chiefly influenced by air temperature.
3. Air temperature affects leaf production, expansion and flowering.
4. Physical and chemical processes within the plants are governed by air temperature.
5. The diffusion rate of gases and liquids changes with temperature.
6. Solubility of different substances is dependent on temperature.
7. Biochemical reactions in crops (double or more with each 10°C rise) are influenced by air temperature.
8. Equilibrium of various systems and compounds is a function of temperature.
9. Temperature effects the stability of enzymatic systems in the plants.

Basic Laws of Temperature

Boyle's Law

For a gas, temperature remaining constant, the product of pressure and volume is also constant, i.e., when volume increases pressure decreases.

Charles Law

The volume of a given mass of gas is directly proportional to the absolute temperature at constant pressure.

Universal Gas Constant

The constant linking pressure, volume, and absolute temperature of a mole of ideal gas, equivalent to 8.3143 joules per kelvin per mole or 1.9858 calories per degree celsius per mole.

Van't Hoff Rule / Q_{10} Rule

The rate of response of a process in crop plants is often doubled or more for each increase of 10°C of temperature within certain limits.

First Law of Thermodynamics

The law of conservation of energy is the first law of thermodynamics which states that in a system of constant mass, energy can neither be created nor destroyed. Mechanical equivalent of heat is the special case of this law.

Temperature Distribution

Each day the earth receives energy in the form of incoming solar radiation from the sun. This shortwave solar radiation ranges mostly from ultra-violet to the near infrared, but, reaches its maximum at around 0.5 microns wavelength (Blue-green visible light). This insolation is absorbed by the earth's surface and is converted to heat (longwave radiation). The earth's (terrestrial) longwave radiation reaches its peak intensity at 10 microns wavelength (thermal infrared) and is responsible for heating the lower atmosphere. There are three temperature distributions viz., horizontal, vertical and periodic distributions that are observed on earth.

I. Horizontal Temperature Distribution

Sun rays make different angles at the same place at different times. Also, different angles at the same time at different places as the axis of the earth makes an angle of 23.5^0 with the vertical. Due to the variation in angle of sun's rays, distribution of solar heat on earth decreases both ways from equator to poles. This is known as horizontal distribution of air temperature.

On maps, the horizontal distribution of temperature is shown by isotherms. The isotherms are imaginary lines drawn connecting points that have equal temperature.

Factors Influencing Horizontal Distribution of Temperature

1. Latitude

The effectiveness of insolation in heating the earth's surface is largely determined by the latitude. So, there is a general decrease in temperatures from the equator to poles, which is a classical example of horizontal temperature distribution.

2. Ocean Currents

Transport of ocean water in the form of currents carries heat from one part of the earth to another which results in horizontal distribution of sea-surface temperature.

3. Mountain Barriers

Mountain Barriers and ranges tend to guide the movement of cold air masses resulting in horizontal temperature variation.

Example : Himalayas protect India from polar air.

4. Topography and Relief

In the northern hemisphere north facing slopes generally receive less insolation than south facing slopes and temperatures are normally lower.

II. Vertical temperature distribution

The decrease of air temperature with altitude is known as vertical temperature distribution.

Example : Permanent snow caps in high mountains.

1. The vertical distribution of temperature is due to adiabatic lapse rate. An adiabatic process is one in which the system being considered does not exchange heat with its environment.
2. The most common atmospheric adiabatic phenomena are those involving the change of air temperature due to change of pressure.
3. If an air mass has its pressure decreased, it will expand and do mechanical work on the surrounding air.
4. If no heat is taken from the surroundings, the energy required to do work is taken from the heat energy of the air mass, resulting in a temperature decrease.
5. When pressure is increased, the work done in the air mass appears as heat, raising its temperature.

6. The rates of adiabatic heating and cooling in the atmosphere are described as lapse rates and are expressed as the change of temperature with height.

7. The adiabatic lapse rate for dry air is very nearly 1°C per 100 metres or 10°C per kilometre.

8. If condensation occurs in the air parcel, latent heat is released, thereby modifying the rate of temperature change.

9. This is known as wet adiabatic lapse rate and it is assumed as 0.5°C per 100 m.

10. However, the average adiabatic lapse rate is 6.5°C per kilometre height.

11. Large scale atmospheric motions are approximately adiabatic and clouds and snow or rain associated with them are primarily adiabatic phenomena in that they result from cooling air associated with decreasing pressure of upward air motion.

12. A common example is that of rising "Bubbles" of air on a warm day, leading to cumulus cloud forms, is a simpler adiabatic phenomena on a smaller scale.

13. The growth of such cumulus clouds into thunder clouds is more complex but still a largely adiabatic phenomena.

Primary Causes for the Fall in Temperature with Higher Altitudes

1. The major source of heat for the air at the ground level is the earth. Clearly, with increasing distance from the source of its heat the air temperature must decline.

2. The density of water vapour decreases with elevation so that less heat can be held in the air.

3. Because of lesser pressure at higher altitude the air expands and becomes thinner and cooler resulting in fall in temperature.

III. Periodical Temperature Distribution

The air temperature changes continuously during a day or a year.

(a) Mean Daily Cycle of Air Temperature

1. From sun rise insolation is supplied and the air temperature continuously rises.
2. Maximum air temperature occurs between 1 p.m. and 4 p.m. and minimum temperature occurs just before the sun rise (Figure 3.1).
3. Maximum insolation is received around noon (12 noon) but maximum temperature is recorded from 1 p.m. to 4 p.m. and this delay is known as thermal lag or thermal inertia.

(b) Mean Annual Cycle of Air Temperature

1. The annual temperature changes from one location to other due to many factors.
2. In the northern hemisphere winter minimum occurs in January and summer maximum in July.
3. When loss of longwave radiation exceeds the shortwave radiation received then the temperature falls and under reverse of this situations the temperature increases in a cycle.

Temperature and Climatic Zones

The climatic zones of the earth are :

1. Tropical.
2. Sub-tropical.
3. Temperate.
4. Alpine.

These zones are differentiated on the basis of temperature. The tropical zone is a hot winterless zone; sub-tropical zone is a hot zone with a cool winter; temperate zone is with a warm summer and pronounced winter; and the alpine zone has a short summer and a long severe winter.

Factors Affecting Air Temperature

1. Latitude.
2. Altitude.
3. Distribution of land and water.
4. Ocean currents.
5. Prevailing winds.
6. Cloudiness.

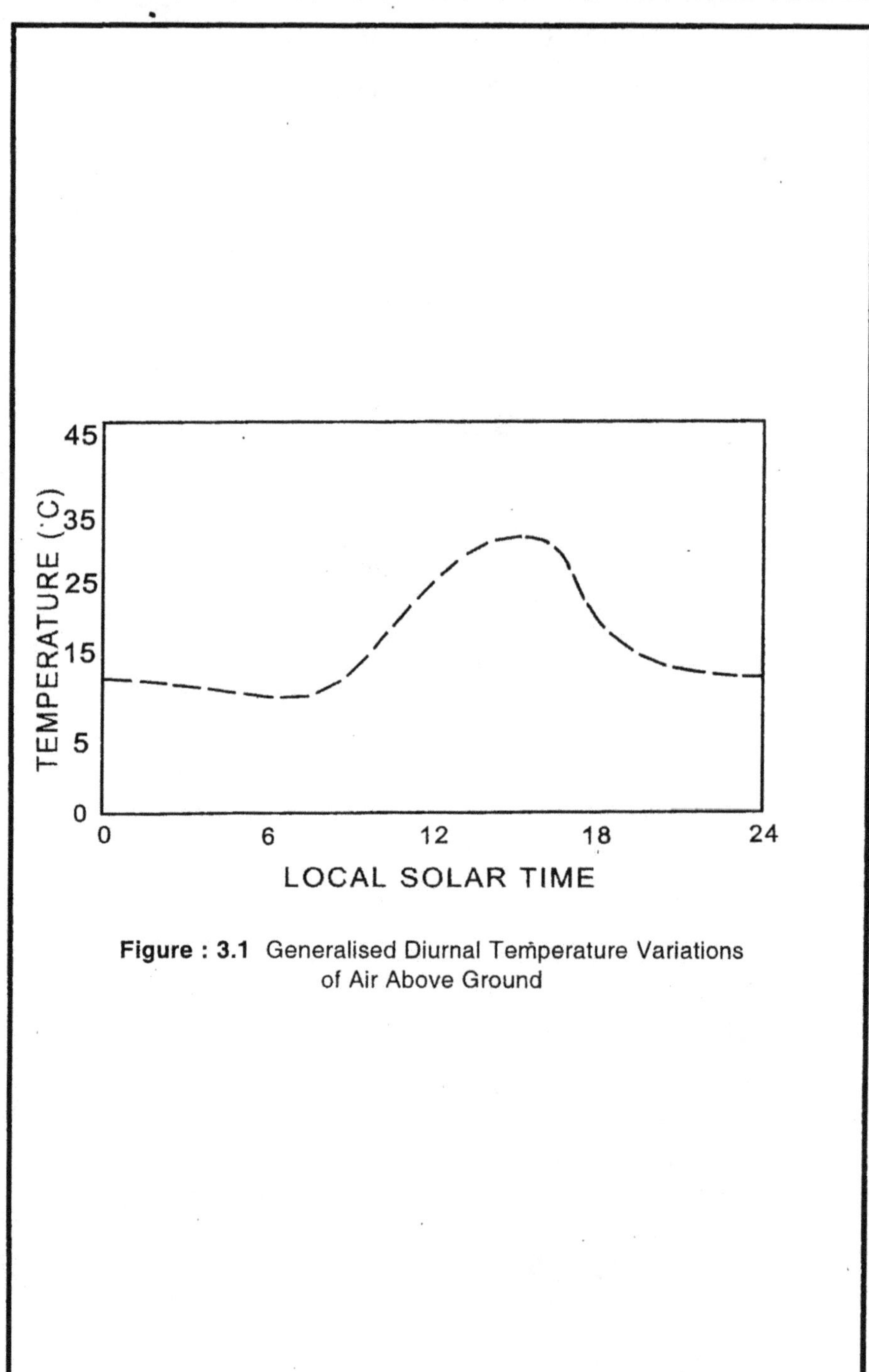

Figure : 3.1 Generalised Diurnal Temperature Variations of Air Above Ground

7. Mountain barriers.
8. Nature of surface.
9. Relief.
10. Convection and turbulence, etc.

Air Temperature and Crop Production

1. Most of the higher plants grow between 0°C and 60°C.
2. The crop plants are further restricted from 10 to 40°C.
3. However, maximum dry matter is produced between 20 and 30°C.
4. At high temperature and high humidity, most of the crop plants are effected by pests and diseases.
5. High night temperature increases respiration and metabolism.

Air Temperature and Plant Injury

The injury to crop plants by air temperature is by both low and high air temperatures.

I. Low Air Temperature and Plant Injury

On exposure of crop plants to low temperature the following effects are observed. The primary effect of low air temperature below their optimum temperature is the reduction of rates of growth and metabolic processes.

1. Suffocation

(a) Small plants may suffer from deficient oxygen when covered with densely packed snow.

(b) Certain toxic substances accumulate in roots and crowns because of low diffusion of carbon-dioxide.

2. Physiological drought

(a) In middle latitudes drought occurs under cool temperature conditions. This is due to excessive transpiration and absence of absorption of moisture from the soil, when the soil is in extremely low temperature conditions.

(b) The internal water content of crop plants is depleted which may result in death of leaves.

3. Heaving

(a) The injury to a plant is caused by lifting upward from the normal position causing the root to stretch or break at a time when the plant is growing.

(b) Sometimes, the roots are pushed completely above the soil surface.

(c) It is difficult for the roots to become firmly established again and the plants may die because of this mechanical damage and dessication.

4. Chilling

(a) Moderate wind speeds when coupled with the air temperature ranging from 0 to 10°C, tends to cause very rapid fall in the activity of metabolic processes, especially respiration in crop plants, which is known as the "Chilling Injury". This results in severe damage and death within a few hours or days.

(b) Due to this injury some crop plants are killed and others recover under favourable conditions later on.

(c) This injury is common in temperate climates where delayed growth and sterility are common symptoms.

(d) Chilling in the affected plants causes a phase change ("Liquid" to "Solid") in membrane lipids resulting in inactivation of membrane bound enzymes.

(e) Sometimes chilling results in yellowing of plants.

5. Freezing

(a) Freezing damage is caused by the formation of ice crystals in the intracellular and extracellular spaces, in the crop plant leaves.

(b) Ice within the cells cause injury by mechanical damage and plant parts or entire plant may be killed or damaged.

(c) If extracellular ice persists, the gradient of water vapour pressure between the apoplast and the cells causes water to migrate out of the cells and into the apoplast, where it freezes, thereby increasing the amount of ice in the plant tissue.

(d) This results not only in mechanical damage to the tissue, but also brings about dehydration of cell contents and lead to death of the cell.

II. High Air Temperature and Plant Injury

1. High air temperature results in the dessication of the crop plants also.
2. The injury caused because of short period fluctuation (within a day highest in noon and lowest at early morning) in air temperature is known as sunclad.
3. The scorching of stem near the soil surface known as stem girdle is another injury at high air temperatures.
4. Plant tissues escape from high heat by emission of longwave radiation, convection of heat and transpiration.
5. However, transpiration is the most effective process in many natural situations.
6. High plant temperatures (> 40°C) are almost invariably due to the cessation of transpirational cooling caused by stomatal closure.
7. Exposure of crop plants to temperatures over 45°C for just 30 minutes can cause severe damage to the leaves of plants.
8. The other effects of high temperature are the disruption of cell metabolism (possibly by protein denaturation), production of toxic substances, and damage to cellular membranes.

Cardinal Temperatures

There are three points of temperature which influences the growth of crop plants. These are termed as the "Cardinal Points" and the synonymous term is the "Cardinal temperature" (Table 3.1).

1. A minimum temperature below which growth ceases (minimum cardinal temperature).
2. An optimum temperature at which the plant growth proceeds rapidly (optimum cardinal temperature).
3. A maximum temperature above which plant growth ceases (maximum cardinal temperature).

TABLE 3.1

CARDINAL TEMPERATURE FOR DIFFERENT CROPS

S. No.	Crop	Min. Cardinal temp. °C	Opt. Cardinal temp. °C	Max. Cardinal temp. °C
1.	Wheat and Barley	0-5	25-31	31-37
2.	Sorghum	15-18	31-36	40-42

Inversion of Temperature

Occasionally at some altitude the temperature abruptly increases instead of decreasing. This condition in which this abrupt rise instead of fall in temperature occurs in the air is known as the temperature inversion. It is due to several reasons.

1. When the air near the ground cools off faster than the over lying layer, because of the heat loss during cooling nights.
2. When an actual warm layer passing over a lower cold layer.
3. Warming by subsidence.
4. Turbulence.

Significance of Temperature Inversion

1. Cloud formation, precipitation, and atmospheric visibility are greatly effected by inversion phenomenon.
2. The upward and downward movement of air currents, mixing of air, etc., are effected by inversion.
3. Impurities like smoke, dust, etc., are confined to lower layers.
4. Fog formation may take place near the ground which may effect the visibility to both human beings and animals.
5. Diurnal (day and night) temperature is also effected by temperature inversions.
6. The incoming solar radiation and its conversion into heat is effected.
7. In the air navigation bumpiness may occur.

Measurement of Air Temperature

Mean and average temperature are calculated by the following methods.

1. Mean temperature of a day $= \frac{\text{Readings at 7 am + 2 pm + 9 am + 9 pm}}{4}$

2. Average temperature of a day $= \frac{\text{Maximum temperature + Minimum temperature}}{2}$

3. Mean monthly temperature $= \frac{\text{Sum of daily means of the month for all days}}{\text{days of a month}}$

4. Mean annual temperature $= \frac{\text{Sum of 12 monthly means}}{12}$

Of the several thermometers used to measure the air temperature for agrometeorological purposes a few are described below.

I. Maximum Thermometer

Principle

The expansion of mercury occurs due to change in the air temperature which is measured by observing the height of mercury column in the bore (capillary) of the thermometer.

Operation and Measurement

Maximum temperature is the highest temperature reached during a particular time in a day. Even though, it may be obtained by observing the thermograph, it is also known from a maximum thermometer, where self-recording thermograph facility is not available. The dry bulb thermometer provides instantaneous values of temperature. But, maximum thermometer is used to measure the maximum temperature (Figure 3.2).

1. The simplest maximum thermometer is a mercury in glass thermometer.
2. The bore in the stem of this thermometer is made extremely fine near the neck of the bulb which is called as 'Constriction'.

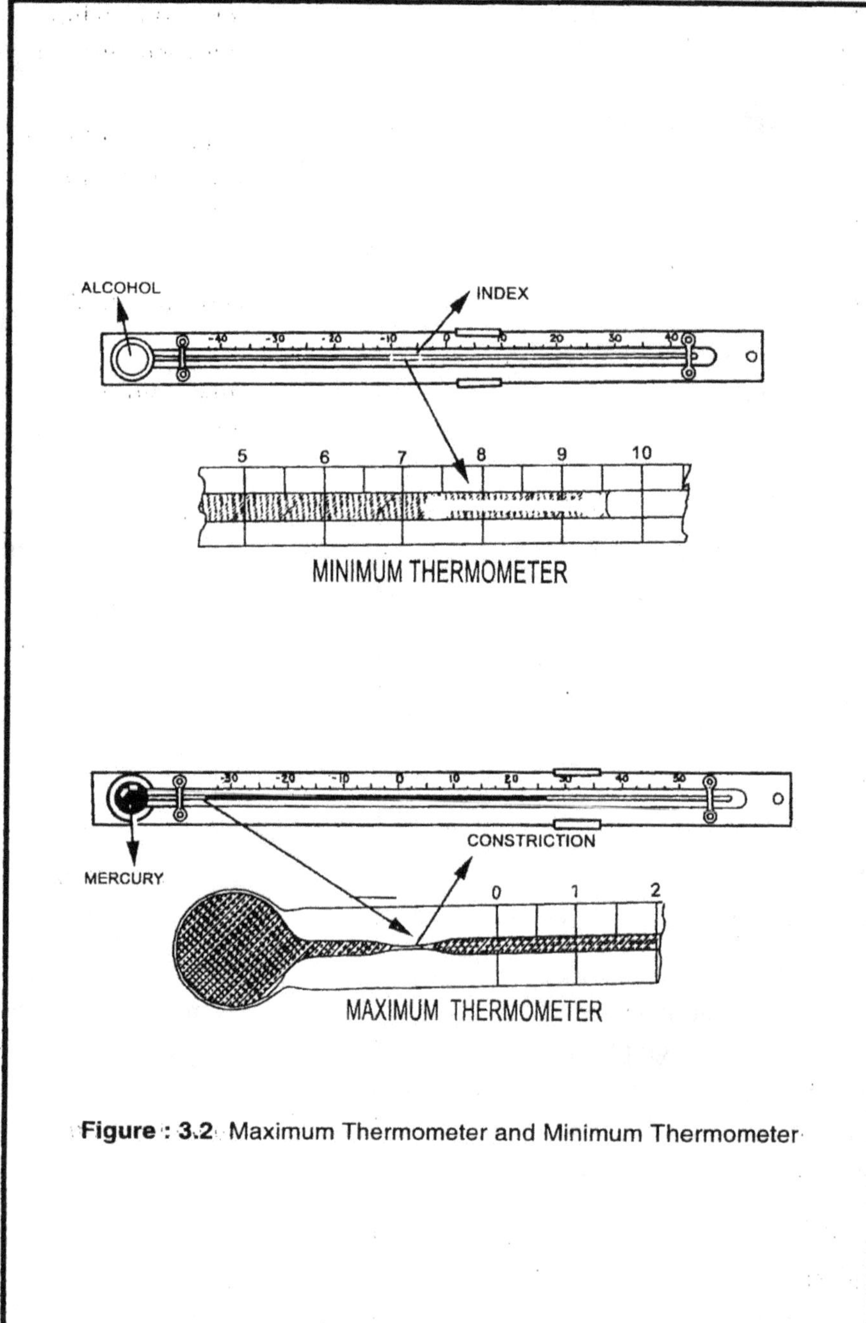

Figure : 3.2 Maximum Thermometer and Minimum Thermometer

3 The presence of this constriction, where the capillary joins the bulb, differentiates this thermometer from other ordinary mercurial thermometers.

4 The constriction is of such a size that, it only allows the expanding mercury to pass, as the temperature rises. But, when the temperature drops, the column of mercury breaks at the constriction leaving a part in the bore.

5 The mercury in the bulb can then contract while the mercury column remains above the constriction.

6. The length of the mercury column in the bore, thus provides the maximum temperature reached since the thermometer was last set.

7. This thermometer is mounted on a special support so that it can be released and turned to a vertical position.

8. This can be reset by whirling or swinging it rapidly in the horizontal direction so that the detached thread of mercury comes down past the constriction into the bulb.

9. Repeat the swinging till the thermometer reads the same value as that of the dry bulb temperature.

10. This should be read to the nearest tenth of a degree.

11. After each observation the maximum thermometer has to be set and kept ready for the next observation.

2. Minimum Thermometer

Principle

The expansion and contraction of alcohol in the bore (capillary) of the thermometer which occur due to changes in the air temperature as recorded by an index.

Operation and Measurement

Minimum temperature is the lowest air temperature recorded during a day or for a fixed time. This may be obtained from the thermograph readings also. However, minimum thermometer provides a record of the lowest temperature occurring at a place of exposure from the time of setting until it is read, where thermograph facility is not available (Figure 3.2).

1. This is an instrument used to measure the minimum temperature of the day.

2. This has a large bore and its fluid is·colourless ethyl alcohol or alcohol.

3. Within the liquid in the bore of the tube, a tiny dark dumb bell shaped index, made up of a metal is present.

4. When the temperature decreases the liquid contracts with the decreasing temperature. The miniscus of the liquid pulls the index down due to surface tension.

5. When the temperature rises again the alcohol flows around the index. The miniscus moves up the bore. However, it leaves the index behind, at the lowest point to which the liquid surface descends to register the lowest temperature reached during day.

6. The position of the end of the dumb bell shaped index, farthest from the bulb (the upper surface of the index) marks the lowest temperature.

7. At the same time the alcohol surface always indicates the current air temperature.

8. The minimum thermometer should always be kept in horizontal position. Otherwise, the metallic index will fall through the liquid to the bottom of the tube.

9. After the readings are taken a magnet is used to reset the thermometer and is restored to the horizontal position. Resetting can also be accomplished by inverting the stem until the index slides down the stem.

10. This thermometer is graduated from -45 to +55 degrees centigrade.

11. This should also be read to the nearest tenth of a degree, like maximum thermometer.

12. After setting the thermometer for next reading, the miniscus of the alcohol should read the same temperature as dry bulb thermometer.

3. Dry Bulb Thermometer

Principle

The expansion or contraction of mercury, as per changes in the air temperature are observed in the bore (capillary) of the thermometer.

Operation and Measurement

1. This is an accurate ordinary mercury thermometer.
2. In this, mercury is enclosed in a sealed glass tube.
3. This has an uniform bore and a bulb at one end.
4. The expansion or contraction of mercury takes place as per the changes in the temperature.
5. This is graduated from -10 to +50 degrees centigrade for direct reading.
6. This instrument always record the current air temperature.

4. Wet Bulb Thermometer

Principle

The expansion and contraction of mercury takes place as per the changes in the air temperature, under wet condition of bulb made moist by muslin cloth.

Operation and Measurement

1. This is also the same as that of dry bulb thermometer. But, the bulb is covered with a muslin.
2. The bulb is continuously kept moist with the muslin cloth.
3. The other end of the cloth is dipped in a container with water; preferably distilled water.
4. The evaporation from the muslin lowers the temperature of the bulb.
5. This thermometer indicates the temperature of the ambient air under saturated conditions.
6. The difference between the readings of wet and dry bulb thermometers is called the wet bulb depression.

7. From the readings of the dry bulb and wet bulb depression, the relative humidity as well as vapour pressure can be found with the help of a calibrated table.
8. The relative humidity is always expressed as percentage.

5. Thermograph

Principle

Two metals having different co-efficient of expansions like invar-bronze are welded together to form a sensitive element. This is connected by a system of linkages to a pen recording on a chart wrapped on a drum which is operated by a clock work.

Operation and Measurement

This is a self-recording device. This is used to obtain continuous and accurate record of air temperature (Figure 3.3).

1. The sensitive element of this instrument is a bimetallic coil as mentioned in the principle above.
2. One end of this element is fixed and the other end touches a delicate set of lever mechanism which operates a pen arm.
3. The tip of the arm is self-inked and touches the calibrated chart wrapped around the rotating drum.
4. The drum completes one rotation in a day of 24 hours and works on a clock mechanism.
5. When the temperature changes the curvature of the bimetallic strip increases or decreases due to difference in the expansion of two metals.
6. These changes are recorded on the calibrated chart.
7. The X-axis of the chart represents time and Y-axis temperature.
8. This device works with a precision of 0.25 degrees celsius.
9. The chart has to be replaced every day.
10. For accurate recordings, the clock should be checked regularly, the recording sheet should be fixed firmly and the pen must be kept clean.
11. The thermograph is located in a double stevenson screen.

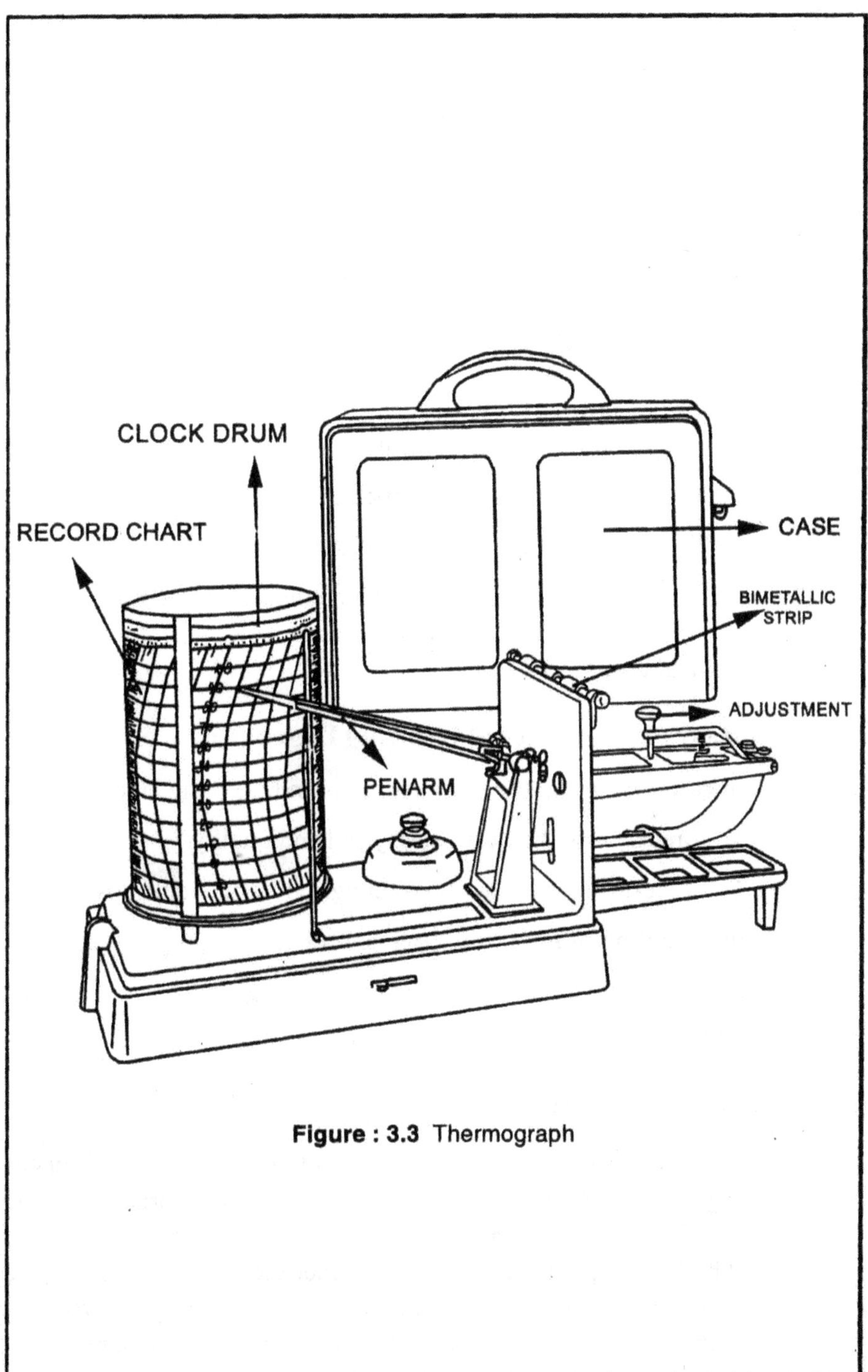

Figure : 3.3 Thermograph

Growing Degree Day Concept

Growing degree day is also known as the "Heat units" or the "Thermal units" or the "Effective heat units" or the "Growth units".

This is a small and simple concept of relating plant growth, development and maturity to the air temperature. The growth of plants is dependent on the total amount of heat to which it is subjected during its life time. The formula for growing degree days is :

$$G.\ D.\ D = \sum_{i=1}^{n} \left[\frac{(T_{max} + T_{min})}{2} \right] - T_b$$

where, T_{max} : Maximum air temperature of the day

T_{min} : Minimum air temperature of the day

T_b : Base temperature of the crop

The base temperature is defined as, "The temperature below which no plant physiological activity takes place". The base temperature for most of the tropical crops is 10°C.

Advantages/Importance of Growing Degree Day Concept

1. In guiding the agricultural operations and planning land use.
2. To forecast crop harvest dates, yield and quality.
3. In forecasting labour required for agricultural operations.
4. Introduction of new crops and new varieties in new areas.
5. In predicting the likelyhood of successful growth of a crop in an area.

SOIL TEMPERATURE

The soil temperature is one of the most important factors that influences the crop growth. The sown seeds, plant roots and micro-organisms live in the soil. The physio-chemical as well as life processes are directly affected by the temperature of the soil. Under the low soil temperature conditions nitrification is inhibited and the intake of water by roots is reduced. In a similar way extreme soil temperatures injures plant and its growth is effected.

Example : On the sunny side, plants are likely to develop faster near a wall that stores and radiates heat. If shaded by the wall, however, the same variety may mature later. In such cases soil temperature is an important factor.

Importance of Soil Temperature on Crop Plants

The soil temperature influences many process.

1. Governs uptake of water, nutrients etc., needed for photosynthesis.
2. Controls soil microbial activities and the optimum range is 18-30°C.
3. Influences the germination of seeds and development of roots.
4. Plays a vital role in mineralization of organic forms of nitrogen (increases with increase in temperature).
5. Influences the presence of organic matter in the soil (more under low soil temperature).
6. Affects the speed of reactions and consequently weathering of minerals.
7. Influences the soil structure (types of clay formed, the exchangeable ions present, etc.).

Variations in Soil Temperature

There are two types of soil temperature variations.

I. Daily variations of soil temperature
II. Seasonal variations of soil temperature

I. Daily Variations of Soil Temperature

1. These variations occur at the surface of the soil.
2. At 5 cm depth the change exceeds 10°C. At 20 cm the change is less and at 80 cm diurnal changes are practically nil (Figure 3.4).
3. On cooler days the changes are smaller due to increased heat capacity as the soils become wetter on these days.
4. On a clear sunny day a bare soil surface is hotter than the air temperature.

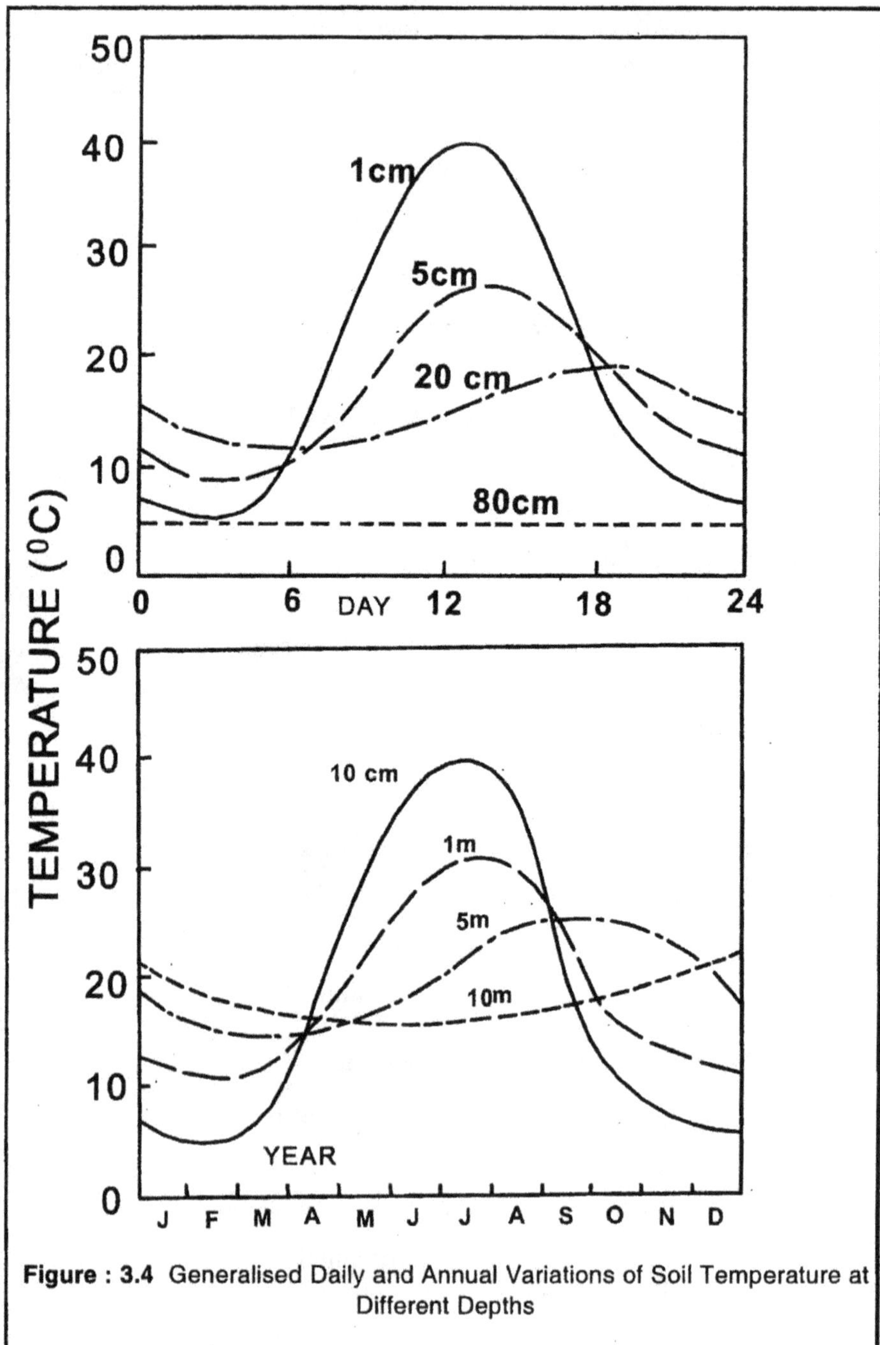

Figure : 3.4 Generalised Daily and Annual Variations of Soil Temperature at Different Depths

5. The time of the peak temperature of the soil reaches earlier than the air temperature due to the lag of the air temperature (Figure 3.5).

6. At around 20 cm in the soil the temperature in the ground reaches peak after the surface reaches its maxima due to more time the heat takes to penetrate the soil. The rate of penetration of heat wave within the soil takes around 3 hours to reach 10 cm depth.

7. The cooling period of the daily cycle of the soil surface temperature is almost double than the warming period.

8. Undesirable daily temperature variations can be minimised by scheduling irrigation.

Seasonal Variations of Soil Temperature

1. Seasonal variations occur much deeper into the soil.
2. When the plant canopy is fully developed the seasonal variations are smaller.
3. In winter the depth to which the soil freezes depends on the duration and severeness of the winter.
4. In summer the soil temperature variations are much more than winter in tropics and sub-tropics (Figure 3.4).

Factors Affecting Soil Temperature

Heat at ground surface is propagated downward in the form of waves. The amplitude decreases with depth. Both meteorological and soil factors contribute in bringing about changes of soil temperature.

I. Meteorological Factors

1. *Solar Radiation*

(a) The amount of solar radiation available at any given location and point of time is directly proportional to soil temperature

(b) Even though a part of total net radiation available is utilised in evapotranspiration and heating the air by radiation (latent heat and sensible heat fluxes) a relatively substantial amount of solar radiation is utilized in heating up of soil (ground heat flux). depending up on the nature of surface.

(c) Radiation from the sky contributes a large amount of heat to the soil in areas where the sun's rays have to penetrate the earth's atmosphere very obliquely.

2. ***Wind***

Air convection or wind is necessary to heat up the soil by conduction from the atmosphere.

Example : The mountain and valley winds influence the soil temperature.

3. ***Evaporation and Condensation***

(a) The greater the rate of evaporation the more the soil is cooled. This is the reason for coolness of moist soil in windy conditions.

(b) On the other hand whenever water vapour from the atmosphere or from other soil depths condenses in the soil it heats up noticeably. Freezing of water generates heat.

4. ***Rain Fall (precipitation)***

Depending on its temperature precipitation can either cool or warm the soil.

II. Soil Factors

1. ***Aspect and Slope***

(a) In the middle and high latitudes of the northern hemisphere the southern slopes receive more insolation per unit area than the northern exposure.

(b) The southwest slopes are usually warmer than the southeast slopes. The reason is that the direct beam of sunshine on the southeast slope occur shortly after prolonged cooling at night, but the evaporation of dew in the morning also requires energy.

2. ***Soil Texture***

(a) Because of lower heat capacity and poor thermal conductivity, sandy soils warm up more rapidly than clay soils. The energy received by it is concentrated mainly in a thin layer resulting in extraordinary rise in temperature.

(b) Radiational cooling at night is greater in light soils than in heavy soils. In the top layer, sand has the greatest temperature range, followed by loam and clay.

(c) The decrease of range with depth is more rapid in light soils than heavy soils when they are dry but slower when they are wet.

(d) Soils with rough surface absorbs more solar radiation than one with a smooth surface.

3. ***Tillage and Tilth***

(a) By loosening the top soil and creating a mulch, tillage reduces the heat flow between the surface and the sub-soil.

(b) Since, the soil mulch has a greater exposure surface than the undisturbed soil and no capillary connection with moist layers below, the cultivated soil dries up quickly by evaporation, but the moisture in the sub-soil underneath the dry mulch is conserved.

(c) In general soil warms up faster than air (Figure 3.5). The diurnal temperature wave of the cultivated soil has a much larger amplitude than that of the uncultivated soil.

(d) The air 2-3 cm above the tilled soil is often hotter (10°C or above) than that over an untilled soil.

(e) At night loosened ground is colder and more liable to frost than the uncultivated soil.

4. ***Organic Matter***

(a) The addition of organic matter to a soil reduces the heat capacity and thermal conductivity. But, the water holding capacity increases.

(b) The absorbtivity of the soil increases because of the dark colour of the organic matter.

(c) At night, the rapid flow of heat from sub-soil by radiation is reduced with the addition of organic matter because of its low thermal conductivity.

(d) The darker the colour, the smaller the fraction of reflected radiation.

(e) The dark soils and moist soils reflect less than the light coloured and dry soils.

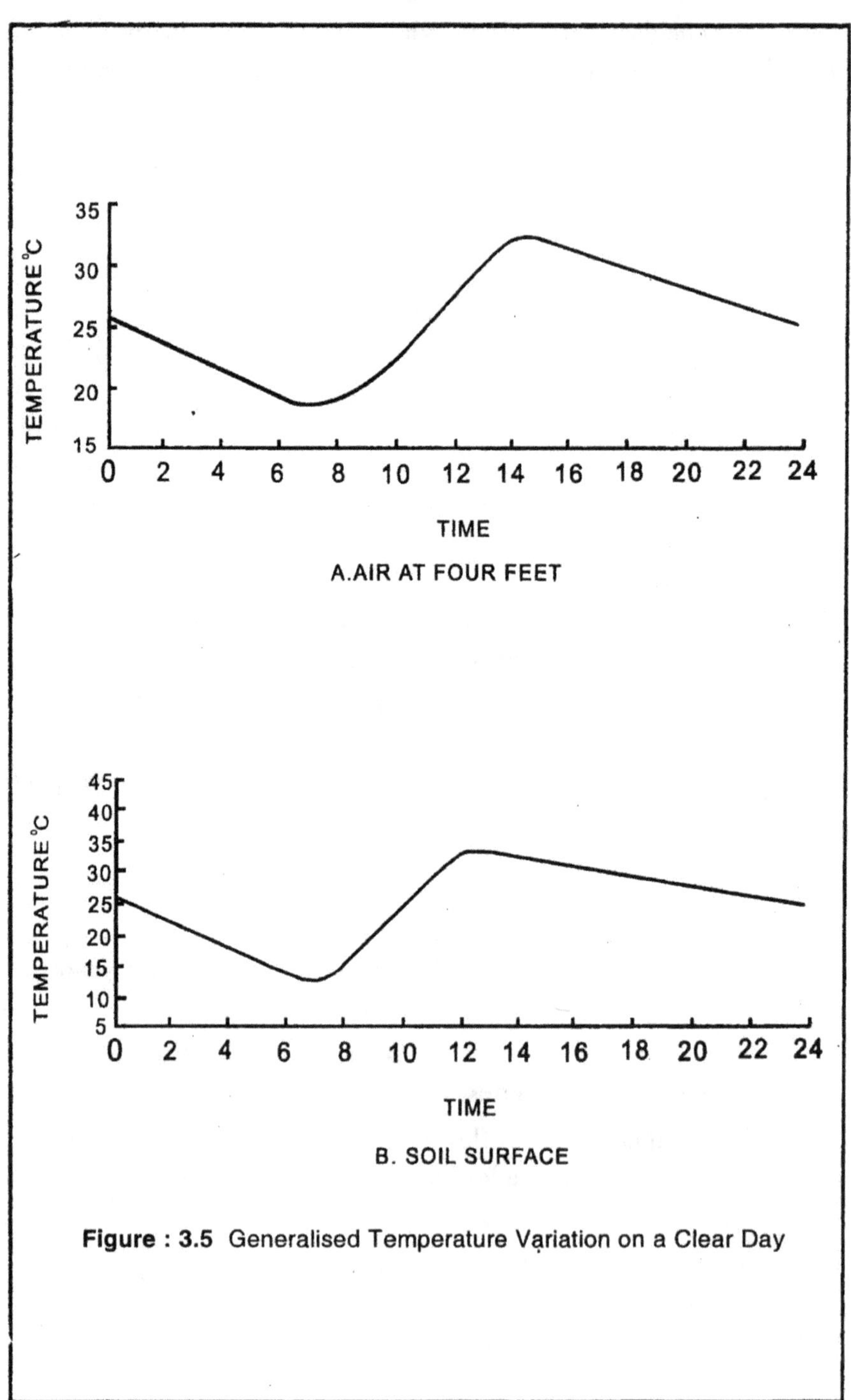

Figure : 3.5 Generalised Temperature Variation on a Clear Day

5. Soil Moisture

(a) Moisture has an effect on heat capacity and heat conductivity.

(b) Moisture at the soil surface cools the soil through evaporation.

(c) Therefore, a moist soil will not heat up as much as a dry one.

(d) Moist soil is more uniform in temperature throughout its depth. as it is a better conductor of heat than the dry soil.

Measurement of Soil Temperature

Soil Thermometer

Principle

Same as mercury in glass thermometers.

Operation and measurement

1. For measurement of earth temperature at shallow depths, the thermometers are bent in between 60 and 120° for convenience of reading (Figure 3.6).
2. Even though the vessel is arranged in the corresponding depth, its scale is exposed for easy view.
3. Mercury in glass thermometers are also used for depths of 50, 100 and 150 centimetres.
4. These thermometers also consist of glass vessel filled with mercury and of the glass capillary fused to it.
5. As soon as the temperature rises, the mercury is pressed into the glass capillary.
6. A scale support is arranged behind the capillary. Scale support and capillary are surrounded by an enclosing tube, the lower part of which is melted together with the mercury vessel.
7. The immersion depth is calculated from the lower end of the thermometer bulb into the middle of the bulb situated in the upper part of the stem.
8. In some commercial makes the bulbs are embedded in wax and hung in metal tubes in contact with soil. Care is also taken to see that water does not enter the tubes.

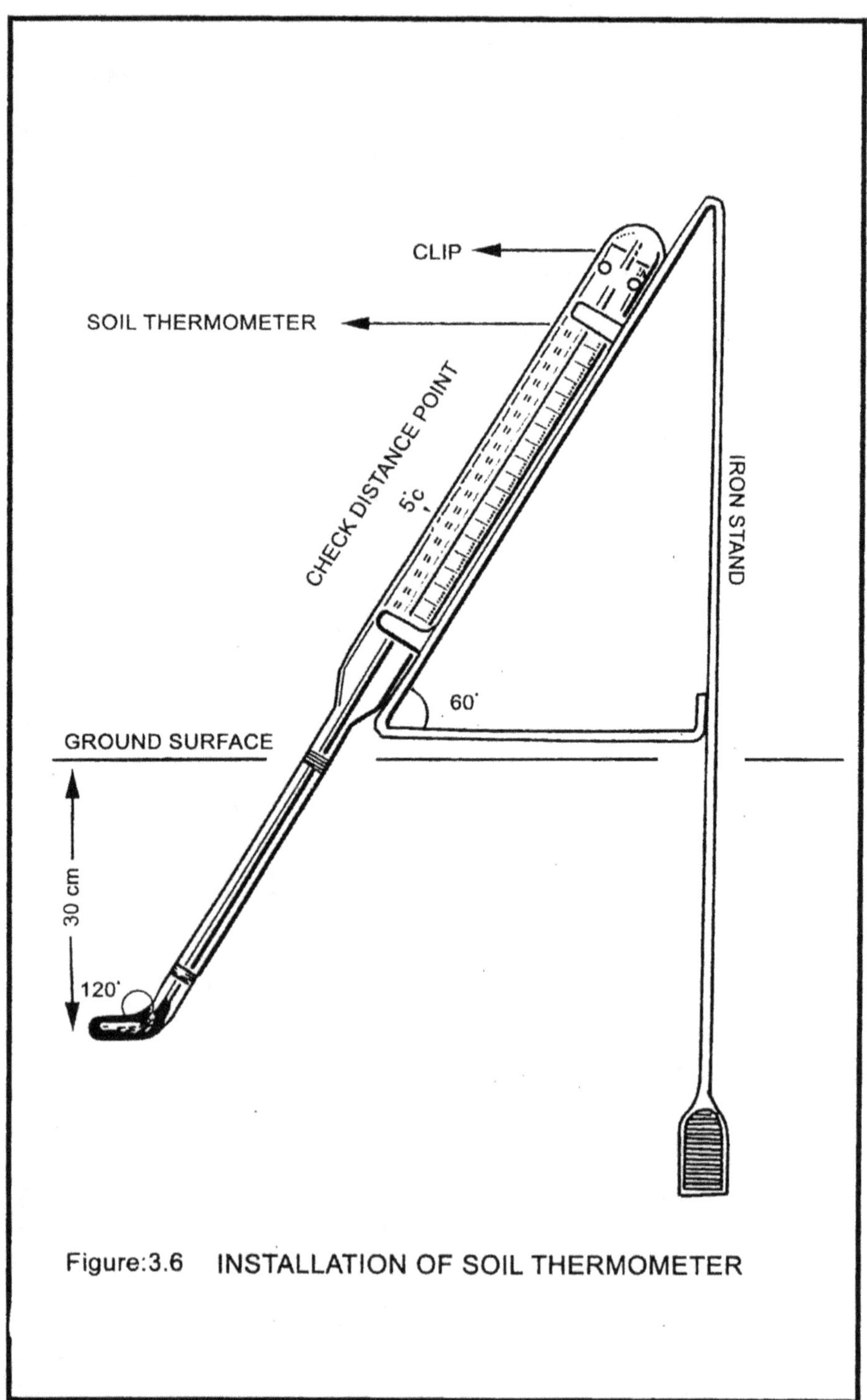

Figure:3.6 INSTALLATION OF SOIL THERMOMETER

9. In all the observatories, the earth thermometers are installed on a level bare ground with no water logging even during heavy rains.
10. It is desirable to keep each thermometer 45 centimetres apart, and the scale portion of each thermometer shall be inclined towards the North. The measuring field should be free from weeds.
11. These are also read at the two usual prescribed timings. The thermometers 100 and 150 centimetres depths are to be removed for a while every time from their places to record the observations.

- *Note :* The above procedure is to be followed in the measurement of earth temperature in an observatory but not in a cropped field.

Basic Temperature Terminology

Basic Thermal Properties

Thermal Conduction

(a) ***In a Gas***

In a gas the phenomenon of thermal conductivity is explained like viscosity. In a column of air the energy varies from layer to layer. This clearly indicates that the thermal conductivity is due to the 'transport of energy'. While conducting heat, the heat causes an equality of temperature through the whole mass of gas under consideration. This is the phenomenon of thermal conduction in a gas.

(b) ***In a Solid***

Thermal conduction in a solid is measured by stating the thermal conductivity, which in turn is the ratio of the steady state heat flow along a long rod to the temperature gradient along the same rod. This varies among different types of solids and depends on temperature and purity of solid.

Thermal Capacity (Heat Capacity)

The quantity of heat in Joules necessary to raise the temperature of a cubic centimeter body or system through one degree kelvin.

Thermal Diffusivity

The thermal conductivity divided by the product of the density and specific heat capacity of a substance. Thermal diffusivity is a measure of facility with which a substance will undergo temperature change.

Specific Heat

The ratio of the quantity of heat required to raise one gram of a substance by one degree centigrade to that required by one gram of water to raise one degree centigrade.

Ground Heat Flux

The rate of transfer of heat to and from the surface of the soil. In the crop radiation energy studies, this is considered as the amount of heat energy expended in heating the soil. This is measured with heat flux plates.

Sensible Heat Flux

This is the product of heat capacity times the kelvin temperature, at constant pressure for a perfect gas. Also known as enthalpy. In crop canopies the heat energy utilised in raising the temperature of air is referred to as sensible heat.

Latent Heat Flux

The amount of heat energy utilised in evaporation and transpiration in a crop canopy.

Results of Research on Effect of Temperature on Important Crops

Rice

Temperature regime generally influences not only the growth, but also the growth pattern. During the growing season, the mean temperature, the temperature sum, range, distribution pattern and diurnal changes, or a combination of these may be highly correlated with grain yields.

A. *Effects of Low Air Temperature*

1. A variety which is susceptible to low temperature when held at 15°C for 4 days during reduction division stage, the sterility percentage was 51. Whereas a low temperature tolerant variety recorded only 5 per cent sterility.

2. Temperature as low as 12°C will not induce sterility if they last for 2 days, but will induce about 100 per cent sterility if they last for 6 days.
3. Injury due to low temperatures is a major constraint in rice production in hilly areas, temperate regions and sub-tropics. For example, the hilly regions of Northern India have about 2.5 m. ha. of rice land spread over Kashmir, Himachal Pradesh, Manipur, Meghalaya, Assam, Arunachal Pradesh, etc., and this injury is a common feature.
4. In temperate regions, cold injury is the main constraint limiting the rice growing area and length of the growing season. The major factors that cause cold injury to rice are (a) cool weather and (b) cold irrigation water.
5. The common types of symptoms caused by low temperature are :

   (a) Poor germination.
   (b) Slow growth and discolouration of seedlings.
   (c) Stunted vegetative growth characterized by reduced height and tillering.
   (d) Delayed heading.
   (e) Incomplete panicle emergence.
   (f) Prolonged flowering period because of irregular heading.
   (g) Degeneration of spikelets.
   (h) Irregular maturity.

B. ***Effects of High Air Temperature***

1. High temperature is a critical factor in rice grain production. A high percentage of sterility and empty spikelets in rice crop are noticed in oasis areas of Egypt.
2. Several authors reported heading to be a stage at which rice plant is more sensitive to high temperature.
3. It is most common to have maximum daily temperatures from 35-41°C or higher in semi-arid regions and during hot months in tropical Asia.

In these regions a heat susceptible variety may suffer. Generally, the high temperature effects are :

(a) Reproductive – White spikelets, white panicles and reduced spikelet sterility.

(b) Anthesis – Sterility.

(c) Ripening – Reduced grain filling.

4. The occurrence of various phenological events and the biomass production depend on the accumulated heat sums, which is having a strong linear relationship.

Groundnut

Temperature is the dominant factor that controls the rate of growth in groundnut. The diurnal variation of the temperature is more important than seasonal variation on the growth of the plant. The reason is that the soil temperature commonly exceeds 40°C in many parts of the tropics, particularly when the soils are dry.

1. When the temperature of top 10 cm soil is below 18°C the emergence of seedlings is low, thereby the plant population is reduced.

2. If maximum temperature exceeds 54°C the embryo is killed.

3. The rate of growth increases as the temperature increases from 20 to 30°C and maximum growth is seen at 30°C. The optimum temperature for vegetative growth lies between 27 and 30°C (air) depending upon the cultivar. The root and pod rot slowed down when the soil temperatures exceed 25°C and stop at 35°C.

4. The lower critical temperature below which no growth is observed is below 13.3°C. The low soil temperature inhibited nodulation and nitrogen fixation.

5. Temperature also plays a vital role in varying the time taken for initiation and opening of first flower. The high geocarposphere temperature resulted in poor yields and quality.

6. Reproductive growth is found to be great between the temperatures of 24-27°C. But, a constant temperature of about 33°C results in no reproductive growth due to loss of pollen viability.

7. The rate of peg initiation has been found to be increased at the temperatures increasing from 20 to 23°C. The highest pod growth is observed when the temperature range in soil is 30 to 34°C.

8. Soil temperature influences pod growth rate and duration. Under high temperature conditions shrivelled Kernels are seen. Air and soil heat kernels influenced N.P.K. uptake in groundnut.

9. To get good yields the effective heat units required are 1600 from sowing to harvest.

10. There is a positive correlation between temperature and oil content in groundnut. Generally, summer groundnut crop has higher oil content than kharif crop.

Sugarcane

1. A tropical region crop grown between 35°N and 35°S of the equator and demands high temperatures ; and extremes of low temperature effects the early period of growth i.e., formative and elongation phases. The ripening phase is also effected.

2. For the formative phase (90 - 150 days) it requires 24 - 30°C which favour better establishment ; and for elongation phase (150 - 240 days) also same weather conditions are required for elongation of internodes, increase in girth, more number of tiller production and high dry matter accumulation.

3. During ripening period frequent fluctuations in temperature adversely affect the sucrose content (5°C more or less) and cane quality.

4. Growth of sugarcane is often limited when soil temperature falls below 21°C and stops completely below 12°C. The low temperature affects the availability of nutrients like N and P. The chlorophyll content of the leaves is reduced at low soil temperatures.

Cotton

1. Cotton is a tropical plant which needs warmth for its development, and high (36°C) average temperatures are less detrimental to its growth than low temperatures, provided water supplies are adequate.

2. Temperature has a decisive effect on beginning germination, while 18 to 30°C is considered optimum, germination is seriously delayed at

the soil temperature below 18°C; below 14°C cotton seeds does not germinate normally.

3. Delayed germination exposes many seeds to fungus infections and insects, and premature sowing in the spring often causes considerable losses.

Units of Measurement

In agricultural meteorology the two basic scales of temperature used are :

(a) Centigrade (b) Fahrenheit

Relationship between °C and °F is

$°C = 5/9\ (°F - 32)$ $°F = 32 + 9/5\ °C$

In meteorology for upper air measurements the absolute scale of temperature is used. The relationship between °A and °C is

$°A = 273.16 + °C.$

Chapter - 4

Pressure

"The Supreme Lord walks and does not walk. He is far away, but, He is very near as well. He is within everything and yet He is outside of everything".

The earth's atmosphere weighs about 60,000 billion tonnes. It constantly presses upon human bodies and crop plants. At sea level, it exerts a pressure of about 1 kilogram per square centimetre. As human beings, animals, etc., got used to it, the same is seldom felt. The force of gravity attracts molecules of air towards the central mass of the earth, but, collisions between these molecules prevent the atmosphere from falling to the earth. As the air molecules are under gravitational pull of the earth, more molecules are held near the earth than at higher altitudes (Chapter 1). From this, it follows that the higher the altitude the lesser the density of the atmosphere, hence lesser would be the pressure. The concentration of oxygen in air is so low at higher altitudes (at 7000 metres) that the amount of oxygen required by crop plants for respiration is not enough to perform any physiological activity. Due to this problem the crop plants and vegetation, become sparser and sparser as the altitude increases.

Importance of Pressure on Crop Plants

The effects of changes in atmospheric pressure on crop plants are normally studied in controlled environmental conditions, because, the temperature and other environmental elements also change markedly with changing altitude. The lower pressures experienced as altitude increases (Table 4.1) have important consequences for high altitude plant life. At high altitudes and low atmospheric pressures the solubilities of carbon-dioxide and oxygen in water are reduced. This has important consequences for respiration, which relies on the transport of these gases in solution.

1. The plants show stunted growth at higher altitudes as concentrations of oxygen and carbon-dioxide reaches low.
2. At higher altitudes increased ultraviolet rays makes environment less favourable for plant growth.
3. The plants with strong root system and tough stems can live under increased wind speeds at low pressures in high altitude areas.

However, the response of the plants to changes in atmospheric pressures has not received the same attention as that of other weather elements.

Isobar : It is defined as, "The imaginary line drawn connecting the points of equal pressure".

Basic Pressure Laws

Dalton's Law of Partial Pressures

The law of partial pressures states that the total pressure of a mixture of two or more gases or vapour is equal to the sum of the pressures that each component would exert if it were present alone, and occupy the same volume as the whole mixture.

Hydrostatic Equation

The rate of decrease of the atmospheric pressure with height is equal to the product of density of air and acceleration due to gravity. It expresses the equilibrium between the atmospheric pressure and the force of gravity at any point in a non-moving air parcel of the atmosphere.

Pressure Gradient

1. It is defined as, "The decrease of pressure between two points along a line perpendicular to the isobars divided by the distance between the points".
2. The isobars drawn closely on a weather map represent a steep pressure gradient and high velocity winds.
3. The isobars drawn farther apart on a weather map indicate a weak pressure gradient and low velocity winds.
4. So, the pressure gradients are one of the primary causes of air flow or wind, with the direction of flow from high pressure to low pressure areas.
5. In addition to the pressure gradient some other forces like gravity, apparent force due to the earth's rotation, etc., influence the pressure distributions on the earth.

Pressure Variations or Distributions

There are four types of pressure variations or distributions.

I. Horizontal Pressure Variation

1. Temperature and pressure are inverse to each other. Along the equator low pressure (doldrums) exists and in cold polar latitudes 'polar high' exists and these are classical examples of horizontal pressure variations.
2. The horizontal variations of the air pressure on the surface of the earth are indicated by isobars.
3. The horizontal variation of pressure depends on temperature, extent of water vapour, latitude, land-water relationship, etc.

II. Vertical Pressure Variation

The downward force on the atmosphere compresses it so that the greater part of the mass of air is concentrated in the lower layers and the atmosphere declines rapidly with increasing altitude. The changes in the atmospheric pressure with altitude may be seen in Table 4.1. At the highest altitude, higher plant population is of the *crucifer*. This exists at 6300 metres in the Himalayas.

TABLE 4.1

CHANGES OF ATMOSPHERIC PRESSURE WITH ALTITUDE

Altitude (m)	Atmospheric pressure ($x\ 10^5$ pa)
0	1.01
1000	0.90
2000	0.79
3000	0.69
4000	0.62
5000	0.53
6000	0.46
7000	0.38

1. The air near the ground is denser than at certain altitude because of force of gravity on air. This results in a corresponding decrease in the pressure with increasing elevation.
2. The rate at which the pressure decreases with altitude is not constant.
3. The pressure decreases at a much greater rate near the ground than at higher altitude, due to density differences at these levels.
4. However, in the lower part of the atmosphere the rate at which the pressure falls is almost uniform.

III. Diurnal Pressure Variation

Apart from variation of the pressure due to movement of atmosphere the pressure varies diurnally.

1. There is a definite rhythm in the rise and fall of the air pressure in a day.
2. Insolational heating (air expansion) and radiational cooling (air contraction) are the main reasons for diurnal variation in the air pressure.
3. Diurnal variation is more prominent near the equator than at the middle latitudes.

4. There are 2 maxima (10 a.m and 10 p.m) and 2 minima (4 a.m and 4 p.m) of pressure in 24 hours (Figure 4.1).

5. The areas closer to sea level record relatively larger amount of variation than away from sea level.

IV. Seasonal (annual) Pressure Variation

The seasonal variations are more pronounced, particularly, over land masses where heating and cooling of the air has a large effect.

1. Due to the effect caused by annual variation in the amount of insolation distinct seasonal pressure variations occur.
2. These variations are larger in the tropical region than the middle and polar regions.
3. Usually, high pressures are recorded over the continents during the cold season, and over the oceans they are observed during the warm season.

Basic Atmospheric Pressure Patterns

There are various smaller pressure systems closely identified with daily weather changes. These are seen on the daily weather maps.

1. Low pressure systems or cyclones

1. When the isobars are circular or elliptical in shape, and the pressure is lowest at the centre, such a pressure system is called "Low" or "Depression" or "Cyclone".
2. A line of low pressure is called a "Trough" when the isobars are not joined at the ends.
3. The word "Cyclone" is derived from a Greek word "Cyclos" meaning the coils of a snake.
4. In India cyclones occur during the monsoon seasons especially in north-east monsoon.
5. The gales accompanying a cyclone give rise to confused seas, torrential rains and usually approaches the coast at 300 to 500 kilometres per hour.
6. A single severe cyclone can perish hundreds of human lives, animal population and submerge thousands of hectares of standing crop.

7. The diametre of a cyclone ranges from a few hundreds to 2000 kilometres.
8. Floods are the results of the cyclones.
9. The devastation could be attributed to the absence of – Timely warning – Lack of awareness among the people – Inadequate preparedness – Poor response and participation.
10. Cyclones are recurring feature in India.

2. High Pressure Systems or Anticyclones

When isobars are circular, elliptical in shape and the pressure is highest at the centre such a pressure system is called "High" or "Anticyclone". When the isobars are elliptical rather than circular the system is called as a "Ridge" or "Wedge".

TABLE 4.2

DIFFERENCES BETWEEN CYCLONES AND ANTICYCLONES

S. No.	Cyclones	Anticyclones
1.	Lowest pressure at the centre and it increases towards the outer rim gradually.	Highest pressure at the centre and it decreases towards the outer rim gradually.
2.	Relative humidity increases towards centre and bring cloudy weather.	Relative humidity decreases and clouds are dissipated giving fair weather.
3.	Variety of clouds lie at different heights.	Little clouds with cool dry air are usually associated.
4.	Highest rainfall occurs at the front side.	Rainfall is almost negligible.
5.	Wind velocity increases from outer rim to the centre.	Wind velocities are much lesser than cyclones (Wind spirally rushes outward from the centre to peripheri).
6.	Move in anticlock wise in northern hemisphere and clock wise in southern hemisphere.	Move in clock wise in northern hemisphere and anticlock wise in southern hemisphere.

Factors Affecting Pressure of a Place

The atmospheric pressure is highly complicated and varies widely on the globe. There are regions of high and low pressures formed on the earth as a result of the factors mentioned below.

1. Temperature of air.
2. Altitude.
3. Humidity in air.
4. Revolution of the earth.
5. Difference in rate of insolation.

Variations in Air Pressure and Weather

1. All the weather changes are closely related to pressure variations.
2. The standard sea level pressure is 1013.25 mb and it varies from 982 to 1033 mb on the earth.
3. High values of air pressure produce clear and stable weather.
4. Low values of air pressure produce cloudy and unstable weather.
5. A continuously rising air pressure is an indication of clear and stable weather.
6. A steadily falling air pressure is an indication of cloudy and unstable weather.

Measurement of Pressure

On the macro and meso scales of analysis, the measurement of atmospheric pressure is fundamentally important for the analysis of local "Weather". The description of weather fronts and areas of low and high pressure is derived from these measurements. At micro scale the measurement of pressure is not crucial. This is more important in the measurements of gaseous exchanges and for converting measurements at various altitudes to their sea level equivalents.

The atmospheric pressure can be measured by an instrument called the barometer. There are two types of barometers, viz.,

1. Mercury barometers.
2. Aneroid barometers.

Of these two, the most accurate instrument is the mercurial barometer. This is used as standard for calibrating the others.

I. Mercurial Barometers

There are two types of mercurial barometers.

1. Fortin's barometer
2. Kew pattern barometer

1. *Fortin's Barometer*

Principle

Balancing of column of air against a column of mercury in a sealed glass tube. The height of the mercury column is proportional to the pressure.

Operation and Measurement

A. The Fortin's barometer is a familiar sight at most of the micro-meteorological laboratories and is an accurate one (Figure 4.2).

B. It consists of a glass tube of uniform cross section and length, which is closed at one end.

C. It is about one metre in length, filled with mercury and then inverted with its lower end open into a movable cistern of mercury.

D. The cistern vessel contains mercury with a flexible leather bag and screw at its bottom.

E. There are two scales on two sides of the tube, one in centimetres and the other in inches.

F. For accurate readings vernier calipers is also attached.

G. The mercury column in the tube is supported by the pressure of the air on the surface of the mercury in the cistern.

H. To take the pressure reading, the height of mercury column is measured on main scale and then vernier scale is read.

I. To read the Fortin's barometer :

(a) Read the attached thermometer to the nearest degree before the time specified for barometer observation.

(b) Gently tap the cistern and tube of the instrument 2 to 3 times with the fingers.

(c) Raise the surface of the mercury in the cistern by screwing up the plunger at the base, until the tip of the ivory point just touches its image in the clear mercury surface.

(d) Set the lower edge of the vernier tangent to the top of miniscus.

(e) Read the scale and the vernier.

(f) Check the reading by making a fresh setting.

2. ***Kew Pattern Barometer (Fixed)***

Principle

Same as the Fortin's barometer

Operation and Measurement

A. This is also like the Fortin's barometer, but is easy to operate (Figure 4.1).

B. In this instrument the cistern vessel is fixed and has no adjusting screw.

C. To allow the rise and fall of mercury in the cistern the divisions are made unequal.

D. The initial adjustment of cistern is not required.

E. To set and read this barometer :

(a) Observe the thermometer attached to the barometer and note the temperature in degrees absolute.

(b) Tap the instrument gently.

(c) Set the vernier scale properly.

(d) Read the readings.

II. Aneroid Barometers

1. ***Aneroid Barometer***

Principle

Aneroid means without air as also liquid. Use of sylphon cell which is a partially evacuated metal diaphragm, expands or collapses depending upon the outside pressure.

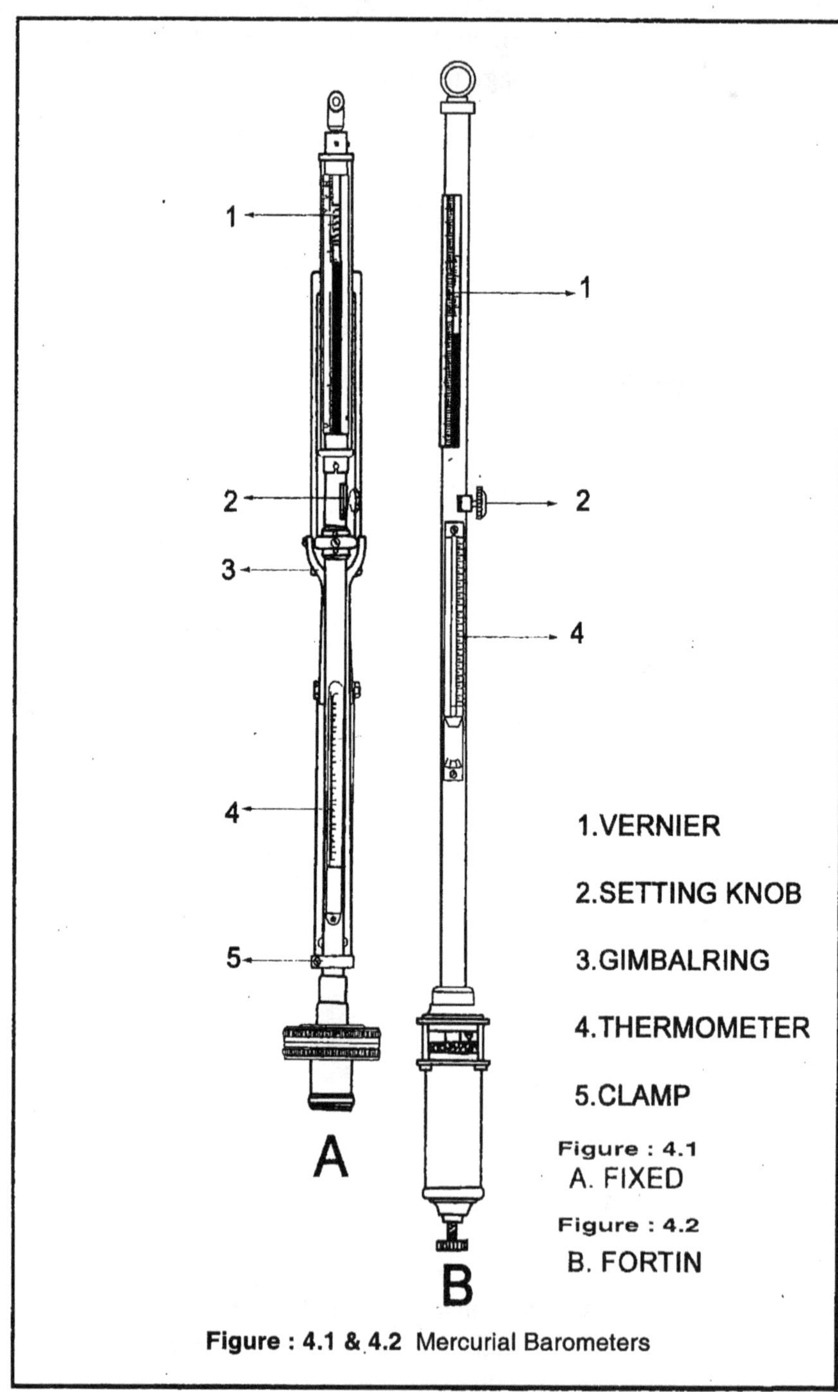

Figure : 4.1 & 4.2 Mercurial Barometers

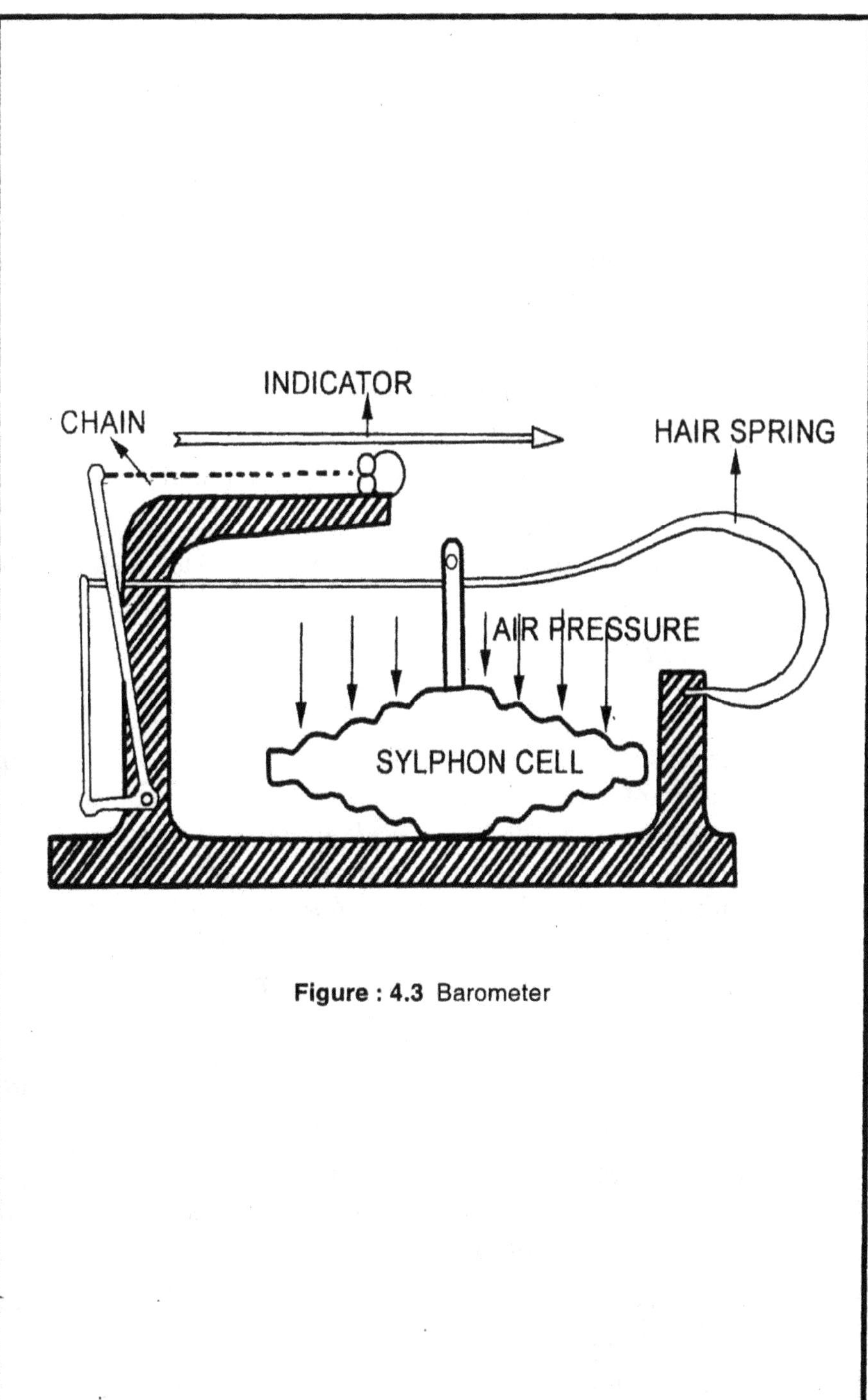

Figure : 4.3 Barometer

Operation and measurement

A. This is constructed with one or more (upto 14) bellows that have been partially evacuated (Figure 4.3).

B. Each bellow may contain an internal spring or may be constructed from tempered steel which acts as a spring.

C. The spring forces the bellows apart against force exerted by the atmospheric pressure.

D. If the atmospheric pressure decreases the springs expand and vice-versa if the pressure increases.

E. The variations of the bellows with pressure changes is mechanically linked to an indicator on a callibrated dial.

F. The aneroid barometers are suitable for outdoor measurements and are also used as altimeters (devices used to find the height above the ground).

2. *Aneroid Barograph*

Principle

The sensitive element in this device is an aneroid capsule which consists of a closed circular vaccum box or boxes placed one above the other. The box is made of an alloy of silver coated berylleium copper. As the atmospheric pressure rises or falls the walls of the box either collapse or distand according to pressure changes. The motion is communicated to a lever system connected to a rotating drum on which recording is made.

Operation and Measurement

This is an instrument used to record the atmospheric pressure continuously.

A. The characteristics of the sensitive element of a barograph are (a) Thin walled (b) Corrugated (c) Silver pliated and (d) Aneroid (Figure 4.4).

B. The movement of the aneroid box corresponding to pressure changes depends upon;

(a) The dimensions of the aneroid box and the thickness of the corrugated diaphragm.

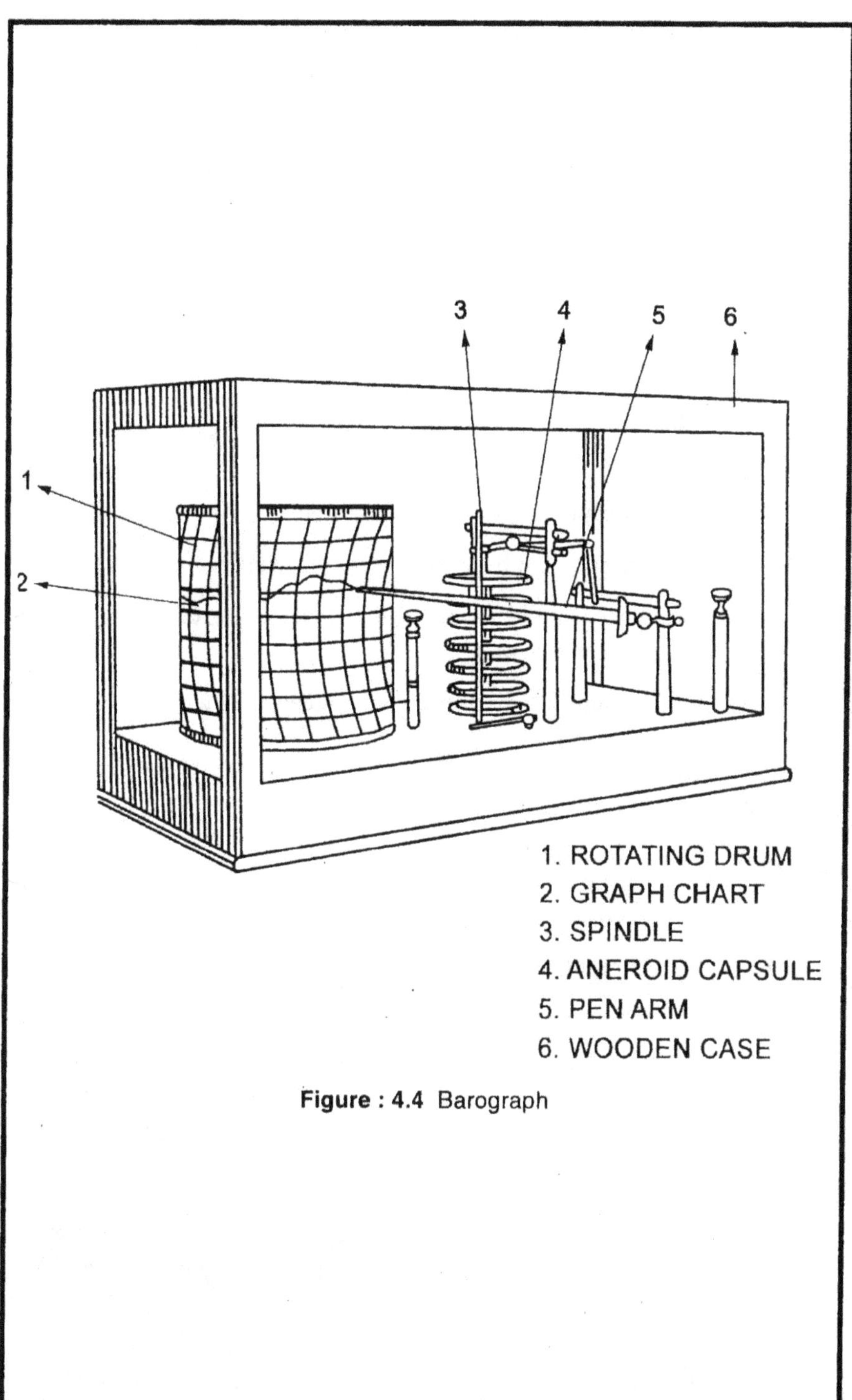

Figure : 4.4 Barograph

(b) The kind of material used in making the diaphragm.

C. The cumulative effect of pressure fluctuations will be recorded by a spindle running centrally through the boxes.

D. One end of the spindle is fixed and the other end touches a lever mechanism.

E. This mechanism operates a pen arm, the tip of which is self-inked and touches a chart wrapped around a rotating drum.

F. This works with a clock mechanism and completes one rotation in 24 hours.

G The graph chart is calibrated vertically in pressure units and horizontally in time units.

H. The pressure is recorded to the precision of one millibar.

I. The chart of the barograph has to be replaced everyday.

J. Any defect due to temperature can be corrected by leaving certain quantity of dry air.

K. To reduce the friction error, a clean pen with proper point should be used.

Basic Pressure Terminology

Pressure

Technically pressure is defined as, "Force per unit area".

Atmospheric Pressure

The atmospheric pressure is defined as, "The pressure exerted by a column of air with a cross sectional area of a given unit i.e., a square inch or a square centimetre extending from the earth surface to the upper most boundary of the atmosphere".

Standard Atmospheric Pressure

The atmospheric pressure varies continuously over a relatively small range and the average of these fluctuations is very close to a value adopted for certain standard conditions defined as "Standard atmosphere". At a temperature of 15°C and at 45° latitude the standard normal pressure is 1013.25 millibars which is equivalent to

29.92 inches (or) 760 mm of mercury at the sea level, which is considered as standard atmospheric pressure.

Units of pressure

1 millibar	=	1/1000th of the bar
1 inch	=	33.86 millibars
1 mb	=	0.75 mm
1 mm	=	1.33 millibar

Units of Measurement

1. Height of the mercury column is measured in inches, centimetres or millimetres.
2. Bar is equal to 10^6 dynes per square metre.
3. The S.I. unit for pressure is Pascal and this is equal to a force of one newton per square metre.

   One atmospheric pressure : 29.92 inches or 76 cm or 760 mm of mercury.

   = 1013.250 millibar.

   = 101.325 kilopascal (kPa).

   = 14.7 lbs/inch^2.

   = 1.014 x 10^6 dynes/cm^2.

Chapter - 5

Wind

"He who sees everything in relation to the Supreme Lord, who sees all living entities as His parts and parcels, and who sees the Supreme Lord within everything never hates anything or any being".

The earth is enveloped by air. The air has weight. So, the pressure exerted by the air is the greatest at the bottom. The pressure decreases with height above the surface of the earth (Chapters 1 and 4).

Air in horizontal motion is known as "Wind". Winds are named by the direction they come from. Windward refers to the direction a wind comes from and leeward is the direction towards which it flows. The wind which flows more frequently from one direction than any other, is called as "Prevailing wind". Wind speed increases rapidly with height above the ground level, as frictional drag decreases (Figure 5.1).

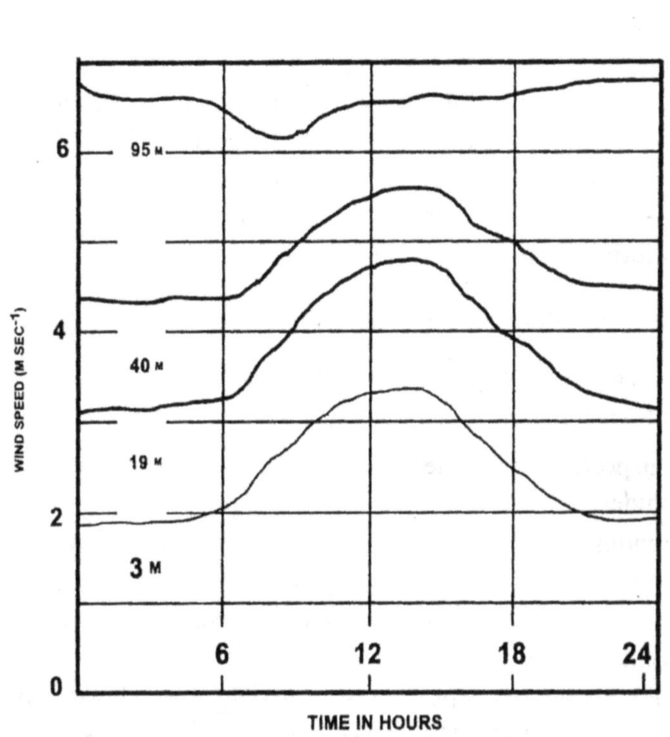

FIGURE: 5.1 GENERALISED DIURNAL VARIATION OF WIND OVER A BARE GROUND.

Usually, the wind can be observed in its undisturbed state only at very high levels. The diurnal variation of the wind speed at various levels (heights) above the ground are as follows.

1. At lower levels there is a daily sequence with a well marked maximum in the hours around noon and a minimum by night.
2. The curve loose this distinct form with increasing height, but, upto 70 metres the trace is apparently uniform.
3. The movement of air at higher levels, where the wind speed depends on the position of the regions of high and low pressure, and not on others. But, this is carried down to lower levels by eddy diffusion.
4. The above figure 5.1 holds good to latitudes affected by migrating high and low pressure areas. In the tropics there is a double variation in diurnal pressure.

Importance of Wind on Crop Plants

Wind performs several functions which are of paramount importance in agricultural activities.

1. Transports heat in either sensible or latent form, from lower to higher latitudes.
2. Provides the moisture (to the land masses) which is necessary for precipitation.
3. Moderate turbulence promotes the consumption of carbon-dioxide in photosynthesis.
4. Wind prevents frost by disrupting a temperature inversion.
5. Wind dispersal of pollen and seeds is natural and necessary for certain agricultural crops, natural vegetation, etc.
6. Action of wind on soil : Wind causes soil erosion in two ways :

   (a) Strong wind blows loose and coarse soil particles (sand) and dust for long distances. In some areas all the soil is blown by this way and no cultivation is possible in such areas.

   (b) In dry countries and sea shores, strong wind is seen to eat up a cliff or a hard rock. When strong wind armed with millions of small particles of sand flows against a cliff or a hard rock, it gradually eats up the rock. The action is strongest near the ground so that the rock is undercut and eventually falls over.

Basic Laws (Essential Features of Kinetic Theory) of Gases

To explain the behaviour of gases, kinetic theory has been very successful. The essential features are :

1. All gases are composed of molecules. The molecules of a particular gas are all alike and differ from the molecules of other gases.
2. The molecules are mere point masses, their dimensions are negligible as compared to the distance between them.
3. The molecules are continuously in motion. They have velocities in all directions ranging from zero to infinity. The velocity of the molecules increases with the temperature.
4. In a steady state the molecules are continuously colliding against each other and also with the walls of the containing vessel. Between two collisions a molecule moves in a straight line and the distance is called "Free path".
5. The collisions are instantaneous.
6. In a direct impact the molecules begin to move in the opposite direction with the same velocity.

Factors Affecting the Wind Motion

1. Pressure Gradient Force

This force is given by pressure decrease per unit horizontal distance and acts in a direction perpendicular to the isobars. This tends to drive the air towards the low pressure.

2. Coriolis Force

This force is due to the rotation of the earth. The coriolis force causes all winds in the northern hemisphere to move towards the right and those of southern hemisphere to move towards the left with respect to the rotation of the earth. At the equator the coriolis force is zero and it is maximum at the poles.

3. Centrifugal Force

When an air particle travels in a curved path, an acceleration towards the centre of rotation is resulted. Due to this acceleration, a centrifugal force operates and makes to throw the particle outward radically.

4. Frictional Force

The roughness of the surface provides frictional resistance to the air motion. This force acts in opposite direction and reduces the wind speed. This force reduces the wind velocity. It reduces geostrophic and centrifugal forces to give resultant in the direction of pressure gradient. The winds are turned towards low pressure and flow across the isobars at an angle of 15° on sea but at 30° on land due to high friction.

Earth's Surface Wind Systems or Wind Belts

The wind belts (Figure 5.2) found on the earth's surface in each hemisphere are :

I. Doldrums.

II. Trade wind belt.

III. Prevailing westerlies.

IV. Polar easterlies.

I. Doldrums

Owing to continuous heating of the earth by insolation, pressures are low and winds converge and rise near the equator. This intertropical convergent zone is known as "Doldrums".

1. These are the equatorial belts of calms and variable winds.
2. The location is 5°S and 5°N latitudes.
3. Wind is light due to negligible pressure gradient.
4. Mostly, there are vertical movements in the atmosphere.
5. The atmosphere is hot and sticky.

II. Trade Winds (Tropical Easterlies)

1. The regular high temperature at the equator results in a high pressure forming in the upper levels over the equator.
2. Then, the air is transferred to the northward and southward directions until 35° north and south in both the hemispheres.
3. Due to this reduction in surface pressure on the equator (doldrums) there is an increase in pressure at 35°N and 35°S which are known as horse latitudes (sub-tropical high).

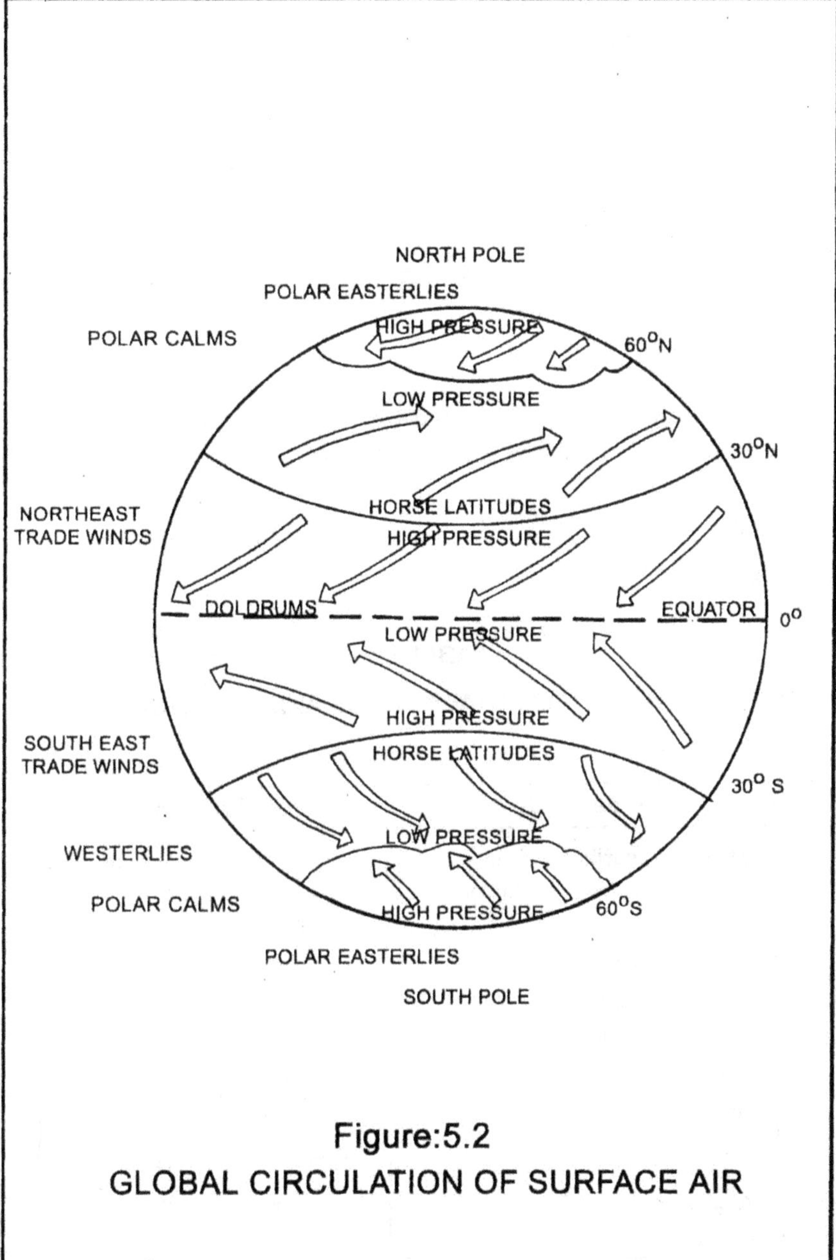

Figure:5.2
GLOBAL CIRCULATION OF SURFACE AIR

4. As a result, the winds flow from the horse latitude to the equatorial region.
5. While moving, these winds are deflected by coriolis force to the right in northern hemisphere and to the left in southern hemisphere.
6. These winds flow from 35°N to the equator in NE direction in the northern hemisphere and from 35°S to the equator in SE direction in the southern hemisphere. These are known as "Trade winds" These are also known as "Tropical easterlies".
7. These are most constant winds in force and direction and flow over nearly half the globe.

Anti - trade winds

1. This is a supplementary wind system of the earth which is effective at higher levels. There are differences between the trade and the anti - trade winds (Table 5.1).
2. This system works in opposite direction to the surface winds.

TABLE 5.1

DIFFERENCES BETWEEN TRADE WINDS AND ANTI TRADE WINDS

S.No.	Trade Winds	Anti Trade Winds
1.	These are surface air currents that move towards equatorial belt.	These are upper air currents occuring above the earth's surface.
2.	Flows from NE in northern hemisphere and from SE in southern hemisphere.	Flows from SW in northern hemisphere and NW in southern hemisphere.
3.	Movement is constant and steady throughout the year.	Movement is not constant throughout the year.
4.	Weather is quiet.	Weather is disturbed.
5	Also known as "Tropical easterlies".	Also known as "Prevailing wetterlies".

III. Prevailing Westerlies

1. The winds that flow from sub - tropical high to the low pressure area at about 60°-70° latitudes in both the hemispheres are known as "Prevailing westerlies".
2. In the northern hemisphere the direction of prevailing westerlies is NW and in the southern hemisphere SW.
3. These winds are forceful and are irregular as compared to the trade winds.

IV. Polar Easterlies or Polar Winds

1. A permanent high pressure exists on the poles.
2. From these high pressures cold winds flow to areas at about 60°-65° latitudes in both the hemispheres.
3. The winds flow in NE direction in the northern hemisphere and in SE direction in the southern hemisphere.

Mountain and Valley Winds

1. The day time up-valley winds and nightly down-valley winds are commonly found in mountaneous regions (Figure 5.3).
2. During day time the slopes of mountains heat up rapidly because of intensive insolation.
3. But, the free atmosphere at the same elevation over the low lands is not heated to the same extent.
4. This results in warm air moving up along the slope. This up slope breeze is called as the "Valley breeze" or the "Valley winds".
5. However, at night the temperature difference between mountain slopes and free atmosphere at the same elevation is reversed.
6. Nocturnal radiation brings about a more rapid cooling of mountain slopes as a result of which the cool air drains into the valley below.
7. This down-slope wind is called the "Mountain breeze" or the "Mountain winds".
8. There are distinctive differences between the mountain and the valley winds (Table 5.2).

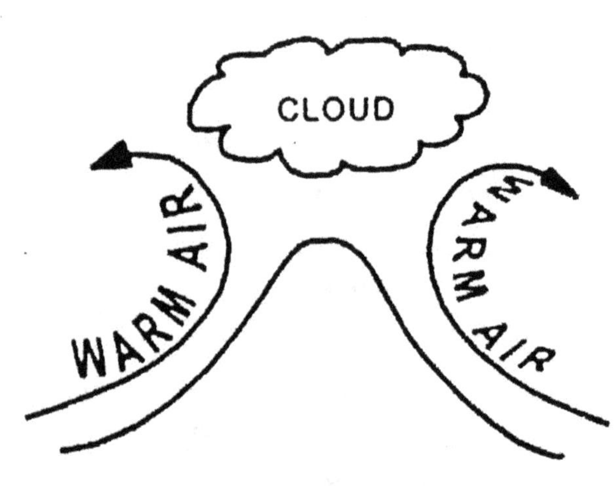

VALLEY WINDS (DAY)

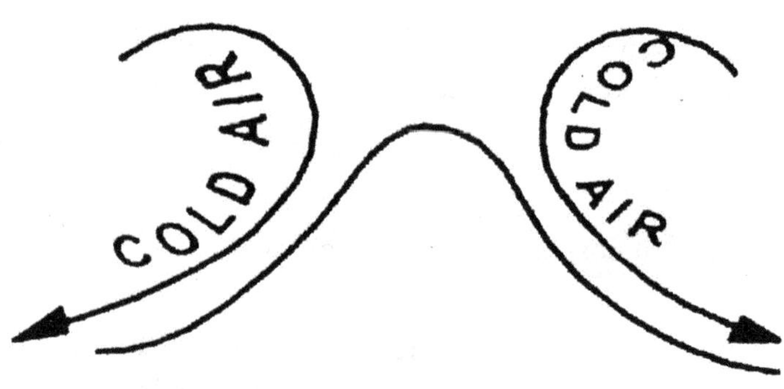

MOUNTAIN WINDS (NIGHT)

Figure : 5.3
MOUNTAIN AND VALLEY WINDS

TABLE 5.2

DIFFERENCES BETWEEN MOUNTAIN AND VALLEY WINDS

S.No.	Mountain winds	Valley winds
1.	Blows from mountain up slope to base.	Blows from valley base to up slope.
2.	Occurs during night time.	Occurs during day time.
3.	Cooling of air close to slope takes place.	Over heating of air adjacent to slope takes place.
4.	Adiabatic heating decreases this phenomenon.	Adiabatic cooling decreases this phenomenon.
5.	Also known as "Katabalic winds".	Also known as "Anabatic winds".

Land and Sea Breezes

1. These winds are defined as, "The complete cycle of diurnal local winds occuring on sea coasts due to differences in surface temperature of land and sea".
2. There is a complete diurnal reversal of wind direction of these coastal winds (Figure 5.4).
3. That is why they are also referred to as diurnal monsoon, since, these wind systems are caused by unequal heating of land and water surfaces.
4. Land and sea breezes are caused by diurnal variation of pressure, but occurrence of monsoons is by seasonal variation.
5. During the day time, more so in summer, land is heated more than the adjacent body of water.
6. As a result warmed air over the land expands producing an area of low pressure.
7. The cooler air over the water starts moving across the coast line from sea to land. This is the "Sea breeze" or "On shore breeze".
8. However, at night because of nocturnal radiation, land is colder than adjacent sea and the pressure gradient is directed from land to sea. There is a gentle flow of wind from land to sea. This "Off-shore" wind is called "Land breeze".
9. There are distinctive differences between the mountain and the valley winds (Table 5.3).

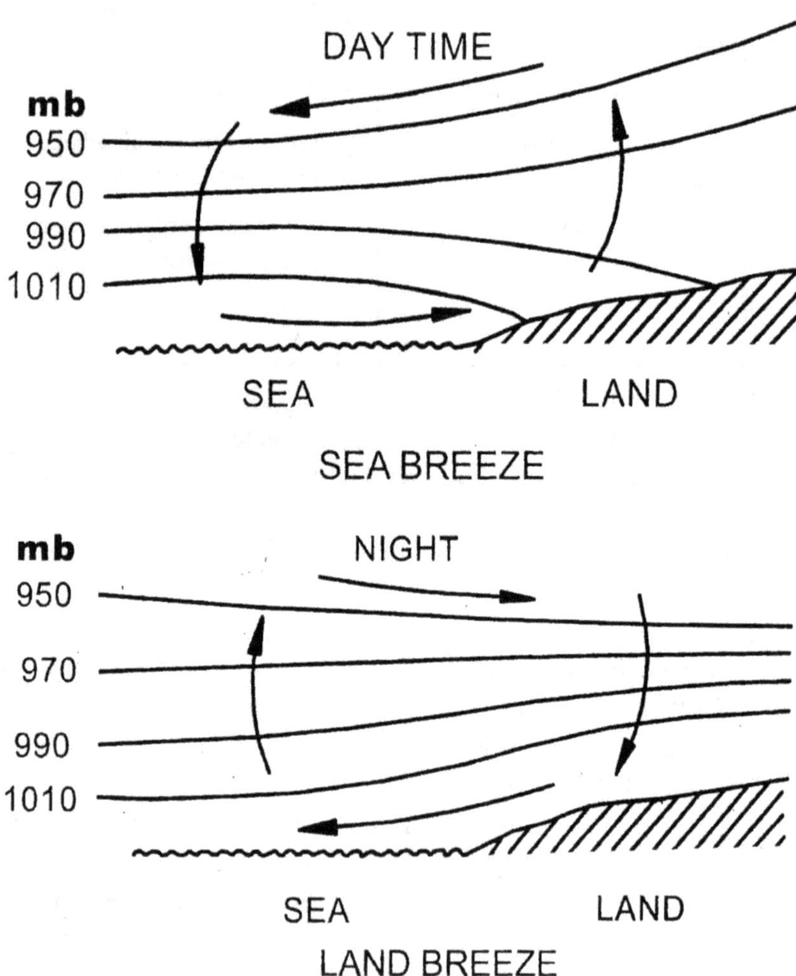

Figure:5.4
LAND AND SEA BREEZES

TABLE 5.3

DIFFERENCES BETWEEN LAND AND SEA BREEZES

S. No.	Sea Breeze	Land Breeze
1.	Occurs in day time.	Occurs in night time.
2.	Flows from sea.	Flows from land.
3.	Have more moisture than land breeze.	Do not have more moisture than sea breeze.
4.	Occurrence depends on topography of coast to a greater extent.	Occurrence depends on topography of land to little extent.
5.	Modifies weather on hot summer afternoons.	Produces cooler winters and warmer summers.
6.	Stronger than land breeze.	Weaker than the sea breeze.

Measurement of Wind

In weather analysis the knowledge of the wind direction and the wind velocity is of great importance. Direction and velocity can be measured accurately by means of instruments. For example, a west wind blows from the west towards the east and a south wind blows from the south towards the north. Wind vane is the most common type of instrument used to determine the wind direction. Wind speed is commonly measured by an instrument called the anemometer. The most common type is the cup anemometer. An instrument called the anemograph makes a continuous record of wind speed.

I. Wind Direction

The direction from which the wind blows is called as the wind direction. This is denoted by two methods. They are (a) In points of compass and (b) Degrees of azimuth, as measured from the true north. The zero point is true north. The other points east, south and west are 90,180 and 270 degrees respectively and these are commonly written as 09,18,27. The North wind has direction 360° (36) and zero (0) is not used except in connection with a 'calm'. In common practice, the wind directions are referred to compass points such as N, NNW,NW, etc.

(a) *Points of Compass*

In this system the four main directions are sub-divided into 8 or 16 and it is called the 8 or 16 point system.

(b) *Degrees of Azimuth*

The zero of the circular scale indicates geographical north, count the degrees while moving in the clock wise direction and it indicates the wind direction. The instruments used to measure the wind direction are :

(a) ***Anemoscope :*** This records the direction of the wind continuously.

(b) ***Aerovane :*** This measures the velocity and direction of the wind instantaneously.

(c) ***Wind Vane :*** This is used in observations to find the wind direction.

Wind Vane

Principle : From the pointer or arrow head of the balancing arm, the indications of the direction are transmitted directly to a cam inside a ring of electrical contacts. By means of contacts corresponding to the cardinal and inter-cardinal points of compass 16 directions can be indicated. Another important means to transmit wind direction (indications) is by means of self-synchronous transmitting motor connected directly to the wind vane.

Operation and Measurement

1. This consists of a balancing arm which is made - up of a very light weight metal or alloy which turns freely about a vertical axis (Figure 5.5).
2. Bearings are provided to minimise frictional losses. The bearings should be good enough to give free turn with light winds.
3. In most common type of wind vanes, one end of the balancing arm exposes a broad surface to the wind. This is bifurcated and is known as fang. While the other end is narrow and points to the direction from which the wind blows. This is known as pointer or arrow head.
4. Under this movable system four to sixteen rods are fixed to a rigid cross.

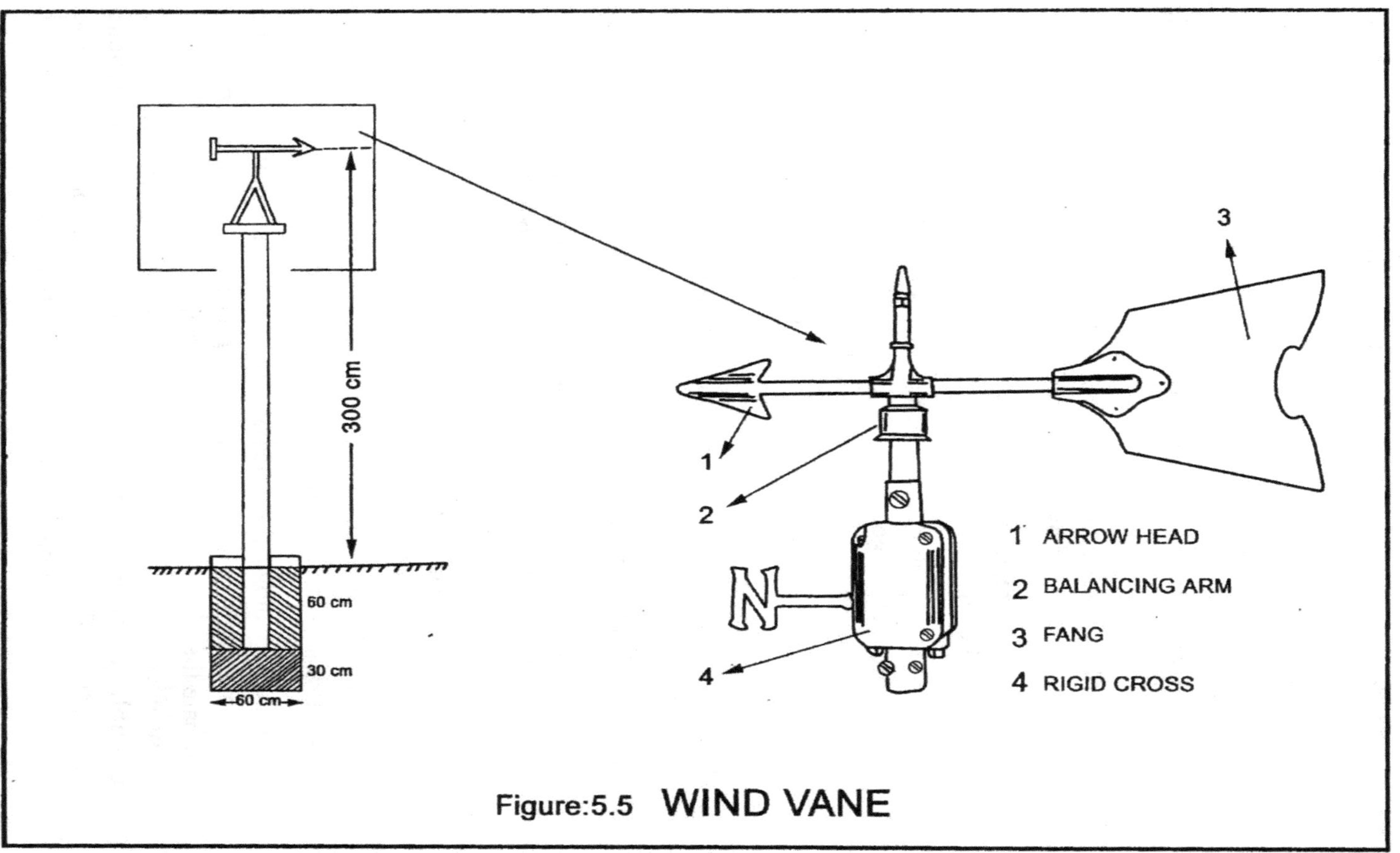

Figure:5.5 WIND VANE

5. The arms of this cross are said to be the four cardinal directions i.e., north, east, south and west.Some other commercial types are provided with eight to sixteen cardinal direction indicators.

6. The wind vane is installed over a wooden plank which is fixed over the wooden post, at ten feet from the ground surface. The north indicator should be set to true north and not to the magnetic north.

7. The observer should stand nearer to the pole and record the mean position of the arrow which oscillates over the cardinal direction.

8. The wind direction should always be recorded as the point from which the wind comes.

9. The wind vane should be watched for few minutes before recording the direction to get the mean observation. The direction of the wind is given by the direction of the arrow.

10. The observer should make sure that the wind vane moves freely on the axis.

11. All parts should be washed with kerosene and lubricated once in every three months.

II. Wind Velocity or Speed

To measure the wind velocity or the wind speed four principal types of anemometers are used in general meteorological work. They are :

1. Rotating cups.
2. Pressure plates.
3. Briddled or torque type.
4. Pressure tube anemometer.

In micro-metrological studies of crop plants, the knowledge of the wind speed with height is necessary for its profile description and to estimate the effectiveness of vertical exchange processes. With the knowledge of the wind speed, at a fixed or reference level, it is also possible to estimate the wind speed at other levels for various applications. From the view point of a micro-meteorologist of crop plants, the anemometers may be classified, as detailed below, depending upon principle used in its manufacturing.

1. Pressure.
2. Mechanical.
3. Thermoelectric.

In the crop micro-meteorological research work, the following are used :

(a) Pressure tube anemometers.

(b) Cup anemometers.

(c) Thermoelectric anemometers.

(d) Sonic anemometers.

The anemometers used in most of the observatories are described below.

1. *Robinson's Cup Anemometer*

This is a rotating cup anemometer, developed in 1846 A.D. This instrument measures the wind speed. Each rotation of the cup wheel corresponds to a definite distance travelled by the wind. Therefore, the number of turns the cup wheel makes in a given time interval corresponds to the distance the wind travelled in that interval. The wind speed can be determined by dividing the distance travelled with time taken.

Principle :

Three or four cups are extended over a vertical axis so that the plane of the cup is in a vertical position. The wind pressure on the concave side of the cup is greater than the convex side. This causes the cups to spin around the vertical axis. By means of proper gear reductions, the rotation of the cups is callibrated in terms of the wind speed.

Operation and measurement

1. This consists of 3 to 4 balancing arms which are made - up of a very light metal or alloy (Figure 5.6).
2. Hemispherical or conical cups are attached to the ends of the arms to provide the necessary pressure gradient which is caused by the convex and concave surfaces of the cup.

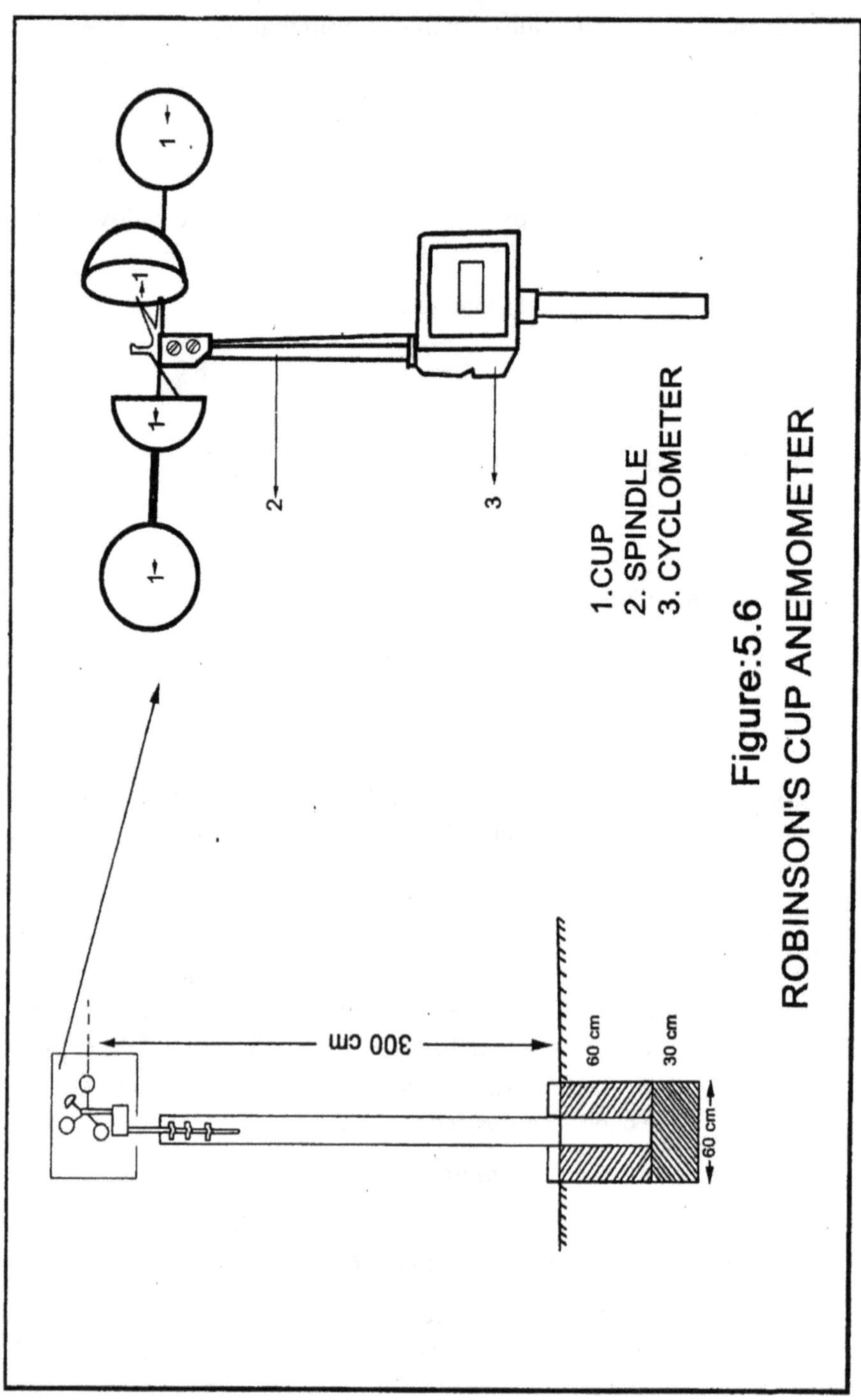

Figure:5.6
ROBINSON'S CUP ANEMOMETER

3. As the force of the wind on the concave side of the cups is greater than that on the convex side, the cups rotate due to kinetic energy.
4. The balancing arm rotates freely over the vertical axis and at the point of articulation, high grade ball barings are provided to minimise frictional losses.
5. The cups are extended on the vertical axis so that the plane of the cup is in a vertical position. The force of the wind causes rotation.
6. The rotating movement of balancing arms is transmitted to the spindle provided in the vertical axis.
7. The spindle is provided with the grooves which operates gauge and this is transmitted to a fine digital meter.
8. Friction is minimised by lubrication and ball bearings. The gauge is calibrated to real units, tenths, hundreds and thousands.
9. The rate of rotation of the cups increases with the wind speed.
10. The box contains a mechanism which establishes a contact when the cups have rotated a certain number of times.
11. The anemometer is kept on a platform at a height of 10 feet from the ground surface and the range of the meter is 0 to 9999.9.
12. An ideal cup anemometer should have no mechanical inertia and starting speed low.
13. In a good cup anemometer there is a linear relationship between the wind speed and the cup rotation per unit time.
14. The wind speed is obtained by measuring the run of the wind in kilometres for a period of 3 minutes at the hour of observation and multiplying it by 20 to obtain the wind speed in kilometres per hour. The mean wind speed in knots is obtained by multiplying the wind in kilometres per hour by 0.54.
15. The wind speed increases with height in accordance with logarithmic law in neutral conditions. Hence, whenever the wind speeds record at different places are to be compared, the height of the anemometer should be taken into account.

16. The height of the anemometer means the height of the cups above the ground.
17. The bearings and gear should be cleaned and lubricated at regular intervals.

2. *Briddled or torque anemometers*

1. These are old type anemometers and can be seen only in old meteorological observatories.
2. These are composed of a series of 32 or more cups around a wheel.
3. The wind force acts on the cups which are held by a spring on the vertical shaft of the cup wheel.
4. The displacement of the cups, or their movement against the spring, represents the wind force.
5. This in turn can be translated into the wind speed.
6. The movement is usually transmitted to an indicating dial or recorded by means of two self-synchronous motors.

3. *Pressure Tube Anemometer*

This instrument is in use from 1892 A.D.

Principle :

The speed of the wind is proportional to the square root of the pressure difference divided by the air density. Since, density changes with changes in pressure and temperature, the speed is obtained from an indicator.

Operation and Measurement

1. The anemometer consists of a "Pilot" tube which is kept pointed into the wind by a vane.
2. The forward end of the tube is open to the wind pressure.
3. It is necessary to measure the pressure difference between the wind pressure and the static pressure or air pressure.
4. A series of holes are arranged around an outer sleeve of the pilot head.
5. Pressure difference is measured by a pressure gauge.
6. Often, the vane and the cups are built into one instrument.

4. *Digital Anemometer*

This is an instrument used to measure the instantaneous speeds of the wind.

Principle :

When the conventionally twisted vane arms are exposed to air, the wind velocity is displayed on a remote sensor. This is done chiefly by a rotating mechanism operated by a battery which is connected to the sensor head.

Operation and Measurement

1. This is a multi-function measurement i.e., m/s, km/h, f/m, and knots.
2. Low friction ball bearings allow free vane movement.
3. Very accurate at both high and low wind velocities.
4. The sensitive balanced vane wheel rotates freely in response to air flow
5. Data hold function helps storing the desired value on display.
6. For easy readability, a large display of low power consumption is provided.
7. Small and light weight design allow one hand operation.

Measurement of Wind Speed at the Hour of Observation

To determine the wind speed at a particular time, two successive readings are taken at an interval of 3 minutes. The difference of the readings is multiplied by 20. For example;

First anemometer reading = 2090.0

Second anemometer reading = 2092.0

Wind speed at that particular hours is

2092.0 - 2090.0 = 2 x 20 = 40.00 kmph.

To determine the average wind speed during the past 24 hours, the readings of yesterday and today are required and the yesterday's reading is subtracted from today's reading. The difference is divided by 24, which gives the average wind velocity for the past 24 hours. For example :

Today's anemometer reading at 08 00 = 9563.5

Yesterday's anemometer reading at 08 00 = 9371.6

The difference = 191.9

Average wind speed = 191.9 / 24 = 8 kmph

Artificial Protection Against Winds

There is a lot of variations (distinctions) between damage to vegetation caused by storms and by winds flowing continuously for long periods. The storm damage is caused by high wind speeds. The gales flatten grain in the standing crops, and forests may up-root and blow down single trees, groups of trees, or whole stands. However, the following few points are of some relevance for consideration.

1. Winds that are moderate to strong, but that flow continuously; as in coastal areas on high plateaus; on mountain tops, have an injurious effect on the water economy of the plants, by placing too heavy demands on it.
2. If the above process continues it adversely affects the water economy, content and other properties of the soil.
3. With increasing intensity of cultivation, especially in years of drought, soil erosion may occur.

Management Through Shelter Belts, Wind Breaks, etc.

An average Indian farmer with enormous natural resources can try for artificial wind protection, which binds the soil, promotes the growth of vegetation and finally increases yields of crops. The protection can be provided through erecting a wind break of non-living materials such as fences of wooden slates, red gram stalks, dried sorghum stems, etc.

Wind breaks provide favourable and unfavourable effects. This is a relative term, which entirely depends on local conditions. For example, a weak wind is usually advantageous for the water balance of the soil, but not in the marsh land which is swampy. Similarly, crops with moist seeds (moisture content more than requirsed for sale or storage) require relatively gentle winds, but not ripe seed, which may show break in seeds.

Benefits of Wind Breaks to Field and Forage Crops and Factors of Shelter Belt Effects on Crop Yields

These can be studied as followed :

1. Land requirement.
2. Competition between shelter belts and crops.
3. Snow distribution in cold areas.
4. Reduction of wind damage.
5. Microclimate modification.
6. Pests, diseases, weed populations, etc.

The variability of shelter belt effects on crop yields

These depend on :

1. Crop.
2. Geographical location.
3. Annual weather conditions.
4. Soils.
5. Shelterbelt design.

Shelter belts have been consistently increased crop yields. The increase in crop yields may be due to the wind erosion prevention and improved micro-climate.

Where the shelter belts are poorly designed the results may not be comparatively encouraging enough. However, the crop yields due to shelter belts can be optimised by :

1. Careful selection of shelter belt species.
2. Shelter belt design.
3. Timely maintenance practices like weed control, etc.
4. Crown - trimming and root pruning may be further advantageous.

Basic Wind Terminology

Windward

The direction from which the wind is blowing.

Leeward

The direction towards which the wind blows .

Veering Wind

The clockwise change in the direction of the wind, which implies that, the wind moves from east to west via south.

Backing Wind

The anti-clockwise direction change of the wind, which implies that, the wind moves from east to west via north.

Logarithmic Wind Profile

A pattern of logarithmic increase of the wind speed with height in the neutral surface layer.

Avogadro's Hypothesis

Equal volumes of two gases at the same temperature and pressure have the same number of molecules.

Results of Research on Effect of Wind on Important Crops

I. Rice

1. A gentle wind during growing period improves grain yield as it increases turbulence in the canopy. The air blown around the plants replenishes the carbon-dioxide supply of the plants.
2. Gentle winds from 0.75 to 2.25 centimetre per second helps in increasing the photosynthesis.
3. However, strong winds (cyclonic) at heading stage may cause lodging. They often dessicate the panicles, increase the floret sterility and increase the number of abortive endosperms.
4. Strong winds enhance spread of bacterial diseases.
5. Dry winds also cause dessication of leaves and mechanical damage to the plants.

II. Groundnut

1. High wind speeds (78 km per hour) result in drying-up of top layer of the soil and do not support germination. It is advisable to go

for deeper sowing at early stages in areas where the high winds and the low relative humidity persist.

2. A frost free gentle wind, during growing period is essential.

III. Sugarcane

1. Since, this crop is sensitive to frost this is a limiting factor in northern India as frost influences the mortality of the seedlings, lower root activity during grand growth period and restricts the growth of the cane.
2. Wind and frost together cause physical damage and splitting of cane thereby solidifies the Juice.

IV. Cotton

1. Apart from appreciably drying-up the soil, moderate winds do little harm, but strong winds and gales seriously affect the delicate young plants.
2. In capsules which open widely on ripening the tips curl outwards and the exposed fibre may be blown away by the wind. Bolls with cup-like form with twisted edges or whose fibre content is compressed are comparatively storm-proof.

Units of Measurement of Wind Velocity

1. Knot
2. Metre per second (m/s)
3. Kilometre per hour (kmph)
4. Miles per hour (mph)
5. Feet per second (f/s).

1 knot = 1 Nautical mile
= 1.15 Status miles/h
= 0.5148 m/s
= 1.853 kmph

1 mph = 0.8684 knot
= 0.447 m/s
= 1.609 kmph
= 1.467 f/s

1 m/s = 1.94 knots
= 2.24 mph
= 3.6 kmph

Chapter - 6

Humidity

"One who always sees all living entities as spiritual sparks, in quality one with the Lord, becomes a true knower of things. What, then, can be illusion or anxiety for him?"

The (invisible) vapour content of the air is known as the humidity. When water vapour changes its physical state appreciable number of heat dynamics takes place resulting in cloud formation on heat loss. When further heat is released the condensed water is changed into droplets and forms either into dew or rainfall.

The humidity affects the rate at which water evaporates from the surface of the plants (transpiration) and this in turn influences the ability of a plant to withstand drought.

Sources of Atmospheric Humidity

1. The water vapour is added into the atmosphere from different water bodies (oceans, rivers, lakes, etc.).

2. Water is transpired into the atmosphere by crop plants, trees, vegetation, etc.
3. Wind carries water vapour into the atmosphere from different places.
4. Through convection and diffusion processes also water vapour enters the atmosphere.
5. The percentage of water vapour is highly variable. It changes according to :
   (a) Season
   (b) Land and sea, etc.

Importance of Humidity on Crop Plants

The humidity is not an independent factor. It is closely related to rainfall, wind and temperature. It plays a significant role in crop production.

1. The humidity determines the crops grown in a given region.
2. It affects the internal water potential of plants.
3. It influences certain physiological phenomena in crop plants including transpiration.
4. The humidity is a major determinant of potential evapotranspiration. So, it determines the water requirement of crops.
5. High humidity reduces irrigation water requirement of crops as the evapotranspiration losses from crops depends on atmospheric humidity.
6. High humidity can prolong the survival of crops under moisture stress. However, very high or very low relative humidity is not condusive to higher yields of crops.
7. There are harmful effects of high humidity. It enhances the growth of some saprophytic and parasitic fungi, bacteria and pests, the growth of which causes extensive damage to crop plants.

   Examples : (a) The blight disease on potato.
   (b) The damage caused by thrips and jassids on several crops.
8. High humidity at grain filling reduces the crop yields.

Terms of Expression of Humidity

The water vapour (humidity) held in the atmosphere at any particular time is expressed in different ways.

I. Mass and Volume Based Terms

1. Specific Humidity

It is defined as, "The mass of water vapour in a sample of moist air to the total mass of the sample". It is expressed as grams of water vapour per kilogram of air.

2. Absolute Humidity

It is defined as, "The actual mass of water vapour present in a given volume of moist air". It is expressed as grams of water vapour per cubic metre or cubic feet.

3. Mixing Ratio

It is defined as, "The ratio of the mass of water vapour contained in a sample of moist air to the mass of dry air". It is expressed as kilogram of water vapour per kilogram of dry air.

II. Saturation Based Terms

1. Relative Humidity (R.H.)

It is defined as, "The ratio between the amount of water vapour present in a given volume the air and the amount of water vapour required for saturation under fixed temperature and pressure". There are no units and this is expressed as percentage.

The R.H. gives only the relative content and indicates the degree of saturation of air. The R.H. of saturated air is 100 per cent.

2. Vapour Pressure Deficit

It is defined as, "The difference between the saturated vapour pressure (SVP) and actual vapour pressure (AVP) at a given temperature".

This is an another measure of moisture in the atmosphere which is useful in crop growth studies. When air contains all the moisture that it can hold to its maximum limit, it is called as saturated air, otherwise it is unsaturated air, at that temperature. The vapour pressure exerted at this temperature under saturated conditions is known as saturated vapour pressure (SVP).

3. Dew point

It is defined as, "The temperature to which a given parcel of air must be cooled at constant pressure and constant water vapour content in order to become saturated".

In this case, the invisible water vapour begins to condense into visible form like water droplets.

Variations in Humidity

1. Absolute humidity is highest at the equator and minimum at the poles.
2. Absolute humidity is minimum at sunrise and maximum in afternoon from 2 to 3 p.m. The diurnal variations are small in desert regions.
3. The relative humidity is maximum at about the sunrise and minimum between 2 to 3 p.m (Figure 6.1).
4. The behaviour of relative humidity differs a lot from absolute humidity. At the equator it is at a maximum of 80 per cent and around 85 per cent at the poles. But, near horse latitudes it is around 70 per cent.

Measurement of Humidity

Water vapour measurement is characterised by a great complexity and range of instrumentation. The instruments used to measure the atmospheric humidity are called "Hygrometers". The three major techniques of measurement suitable for continuous agrometeorological measurement are :

(a) Absorption.

(b) Condensation.

(c) Thermodynamaic.

A few hygrometers developed utilising the above techniques are :

1. Hair hygrometer.
2. Dew cell.
3. Dielectric hygrometer.

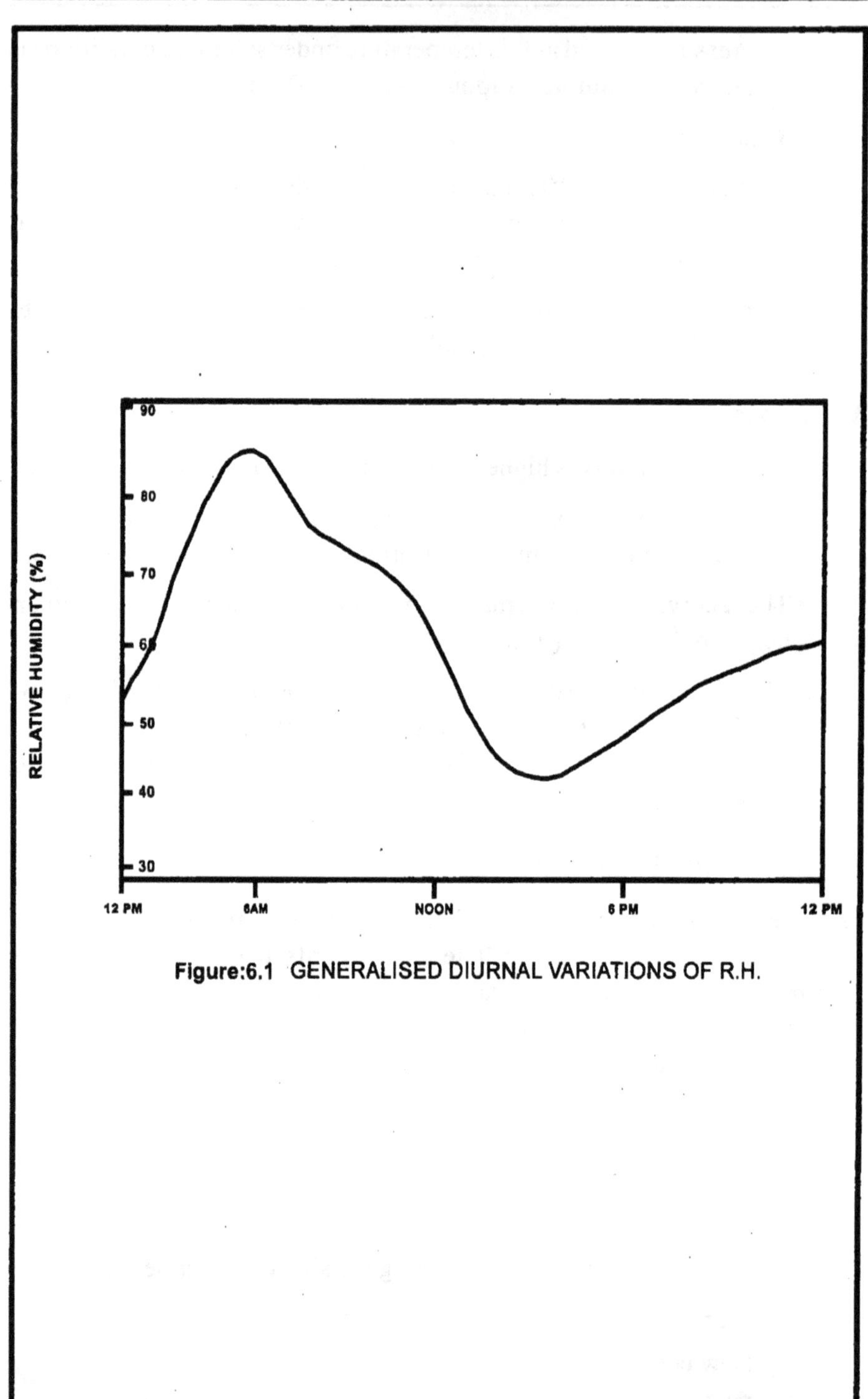

Figure:6.1 GENERALISED DIURNAL VARIATIONS OF R.H.

4. Infrared hygrometer.
5. Wet and dry bulb hygrometer.
6. Hair hygrograph.
7. Sling psychrometer.
8. Assmann psychrometer.
9. Chemical hygrometer.
10. Regnault's hygrometer.

A detailed description of a few hygrometers used in agrometeorological experiments is as follows .

1. Wet and Dry Bulb Hygrometer

Principle : Difference between wet and dry bulb readings indicate the humidity, because of change of rate of evaporation from the underlying wet surface.

Operation and measurement : This hygrometer is handy and an accurate device, and used for outside measurements and also in crop fields.

1. This consists of two mercury in glass thermometers graduated between -10 and 50 degrees centigrade and mounted side by side on a wooden plate (Figure 6.2).
2. One is mounted a little lower than the other and its bulb is covered with a piece of fine muslin cloth, which is wetted with distilled water at the time of observation.
3. The other end of the cloth is dipped in a container filled with pure water (distilled water).
4. Both the thermometers are exposed to air, the humidity of which is to be measured.
5. The dry bulb thermometer indicates the current air temperature whereas the wet bulb thermometer indicates the temperature of air under saturated conditions.
6. The moisture in muslin of wet bulb evaporates and the latent heat is absorbed by the evaporating moisture there by causing the temperature reading of wet bulb thermometer to fall.

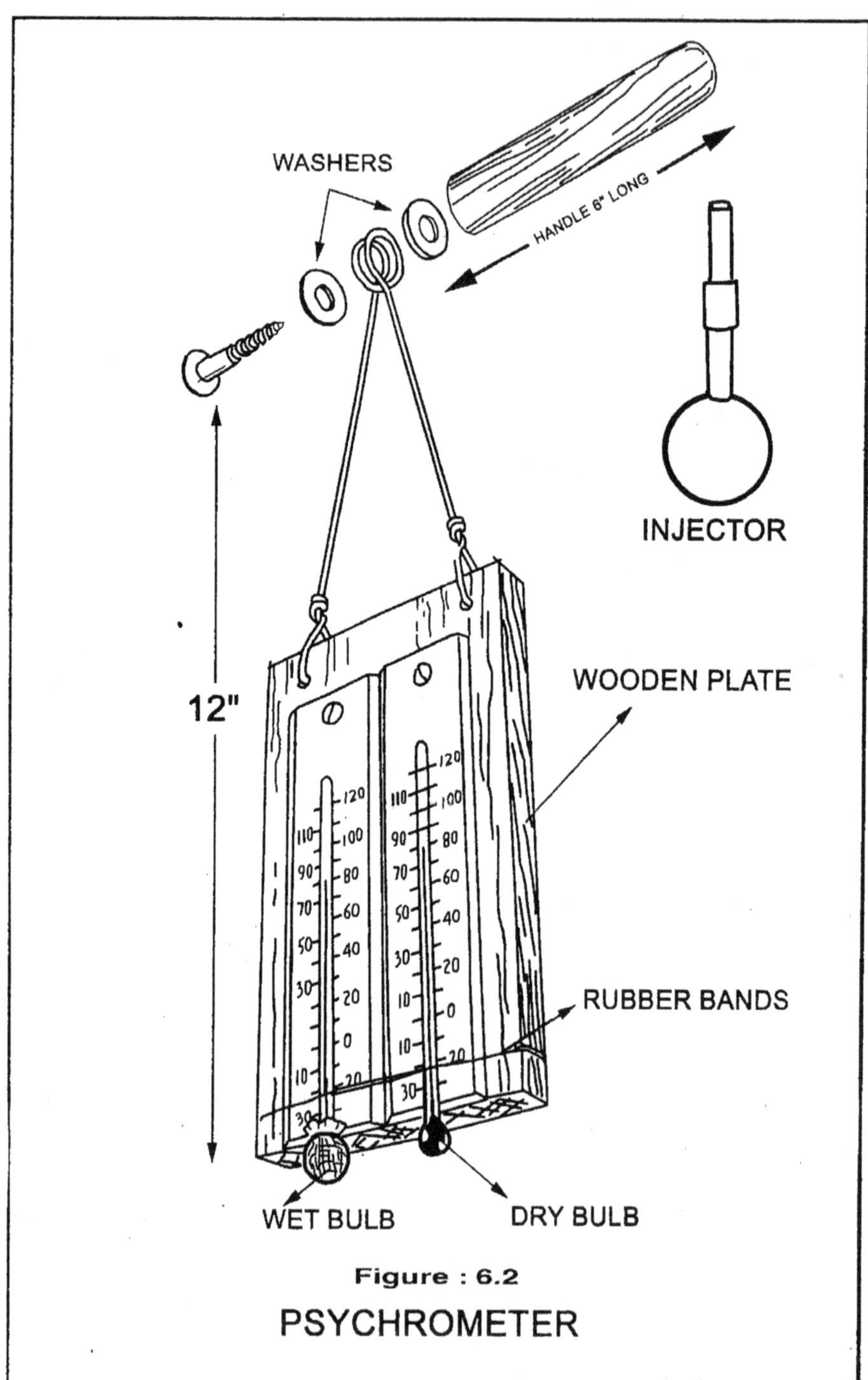

Figure : 6.2

PSYCHROMETER

7. When the speed of the unsaturated air is equal or more than five feet per second and passes over the bulb of the thermometer, water evaporates from the wet bulb depending upon relative humidity and rate of air movement.
8. The difference between the readings of the wet and the dry bulb thermometers is called "Wet bulb depression".
9. From the readings of the dry bulb and the wet bulb depression, the relative humidity is found with the help of calibrated tables. The relative humidity is expressed as percentage.

   Note :

   1. The relative humidity can also be measured from the readings of the dry and the wet bulb temperatures recorded in the Stevenson Screen using psychometric charts. A wet and dry bulb thermometer differs from the thermometers mounted in Stevenson Screen. In Stevenson screen the thermometers are mounted seperately, apart from each other, where as in case of the wet and the dry bulb thermometers, they are mounted side by side on a frame, being protected from exposure to direct solar radiation.
   2. A psychrometer is a pair of thermometers one of which is covered with a wet sleeve. Atmospheric humidity is uniquely related to the wet and the dry bulb temperatures of a properly ventilated psychrometer. Hygrometers are sensors for measuring the water vapour content of a sample of air. The commercial hygrometers are based on several principles.

2. Assmann Psychrometer

Principle : The difference between the wet and the dry bulb readings when unsaturated air passes over the bulbs of thermometers on forced ventilation is recorded and humidity content is found by using proper tables.

Operation and Measurement : This is the best available psychrometer for humidity measurement in micro-meterological research of crop plants. This is portable and also called as "Aspiration psychrometer".

1. In this instrument two sensitive, calibrated mercury in glass thermometers are enclosed in a double walled radiation screen (Figure 6.3).

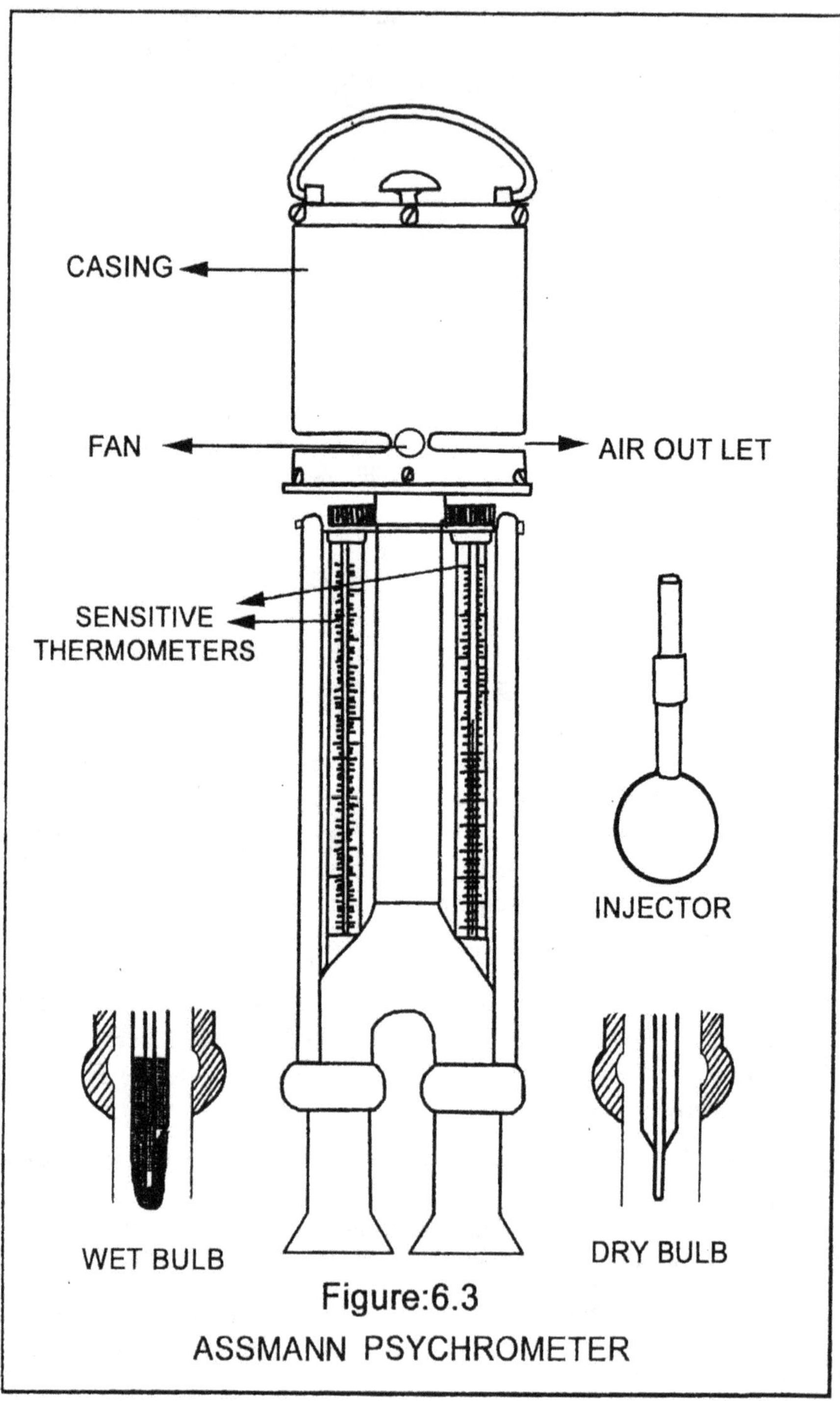

Figure:6.3

ASSMANN PSYCHROMETER

2. Usually, nickel plated coaxial tubes which are thermally insulated from rest of the apparatus are used to minimise the radiation effects.

3. Of the two thermometers, one is a dry bulb and the other one is covered with a thin muslin cap which is moistened with distilled water every time the instrument is used.

4. To ensure adequate opportunity of evaporation from the wick of the wet bulb the psychrometer is aspired by a clock work motor housed in the casing.

5. In the recent commercial versions, a small direct current motor is also being used.

6. The wet bulb must be on the down stream side of the dry bulb i.e., nearer to the fan.

7. This should be done to prevent the cold air from the wet bulb impinging on the dry bulb.

8. The depression of the wet bulb temperature below the dry bulb temperature in an unsaturated air is due to the evaporation of water from the muslin around the wet bulb and the consequent cooling due to evaporation phenomenon. The relative humidity can be computed from the tables.

9. In recent commercial makes the thermometers are replaced by differential thermocouples.

   Note :

   Hygrometric tables meant for the Assmann psychrometer should not be used for computing relative humidity from Stevenson Screen temperatures, and vice versa. Tables appropriate to the altitude of the station should be used.

3. Hair Hygrometer

Principle : When the air is dry, the cells in the hair are close together. But, when the air is humid, the space between the cells absorbs water vapour and the hair thickens and lengthens. This contraction or expansion of hair with change in moisture content is used to measure the moisture.

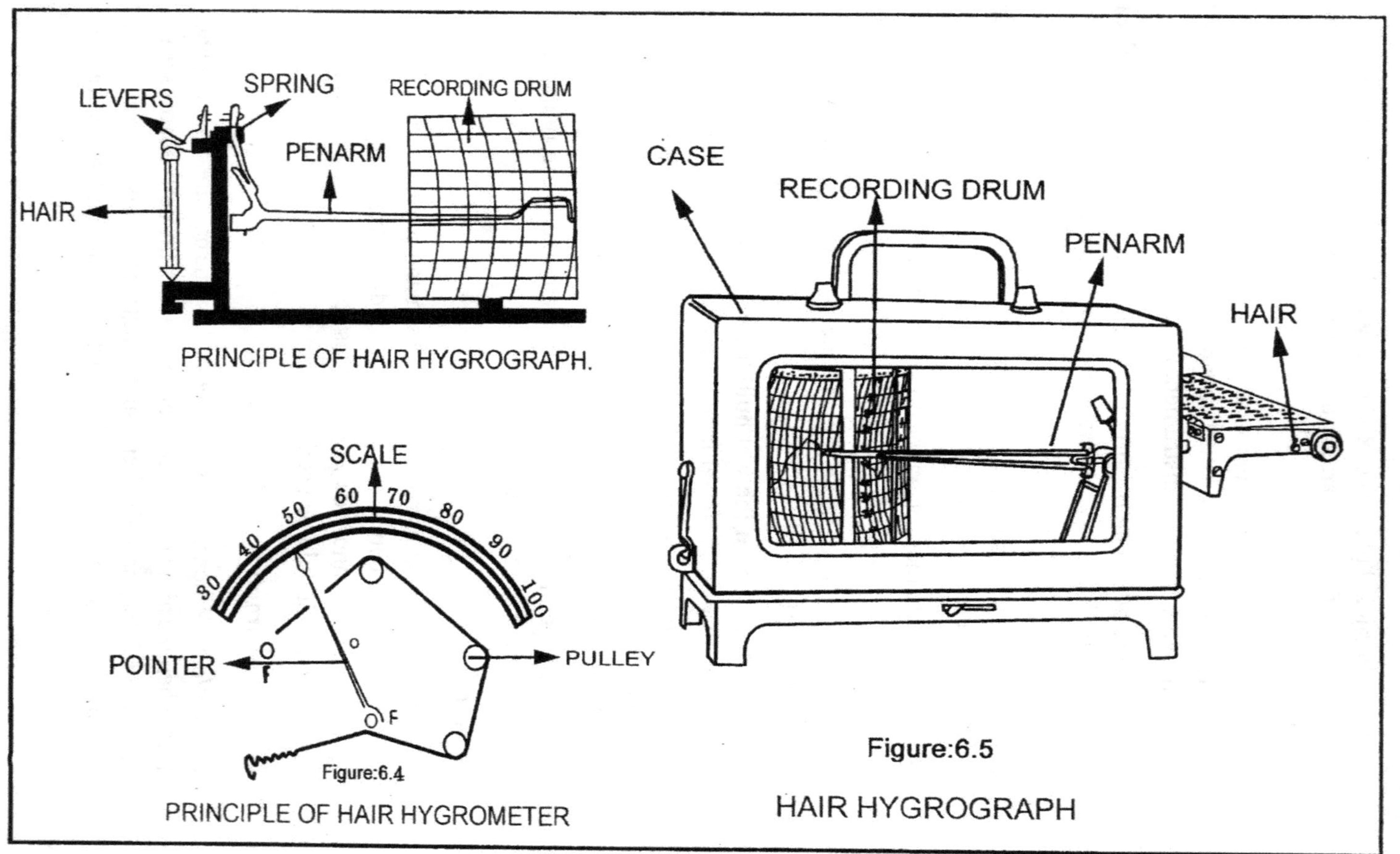

PRINCIPLE OF HAIR HYGROGRAPH.

Figure:6.4

PRINCIPLE OF HAIR HYGROMETER

Figure:6.5

HAIR HYGROGRAPH

Operation and Measurement

1. This instrument measures relative humidity and is easy to carry to distant places.
2. A bunch of human hair is used in a socket.
3. On absorption of moisture, any slight increase in the volume of hair is magnified by a delicate set of lever mechanism (Figure 6.4).
4. To this lever mechanism a pointer is fixed which can move over a scale graduated from 0 to 100 per cent.
5. When the humidity changes, corresponding change occurs in the length of hair.
6. So, the pointer moves because of the movement in the lever mechanism.
7. The pointer which moves across a scale, indicates the relative humidity directly.

4. Hair Hygrograph

Principle : Same as hair hygrometer. However, to record the continuous changes in relative humidity on graph paper during the hours of the day, a recording mechanism is used. When a hygrometer is transformed into a self recording device it is called as a hygrograph. This is used to record the relative humidity of the air continuously.

Operation and Measurement

1. A band of human hair is fixed on the levers and any slight increase in the volume is transmitted to the pen arm.
2. The pen arm is self-inked and works on levers.
3. A change in the length of hair is proportional to the log change of relative humidity.
4. A calibrated chart is wrapped around a rotating drum. This completes one rotation in 24 hours and works on clock mechanism.
5. The X- axis represents time and Y-axis, the relative humidity.
6. The chart has to be replaced everyday.
7. The dust on the hair should be cleaned and washed regularly.

8. The hair should not be touched with hand. This instrument should be exposed in double Stevenson Screen.
9. The screen should be located in a place where the air is not polluted with smoke, dust, oil and ammonia releasing industries in the immediate surroundings.

Units of Measurement

Percentage (%)

5. Stevenson Screen

1. It is a wooden box designed by a British scientist Mr. Stevenson.
2. It is made-up of small pieces of wood (panes) which are fitted obliquely to prevent direct entry of air inside (Figure 6.6).
3. It is provided with double roofing with air in between the two roofs.
4. As air is a poor conductor of heat, it serves as an insulator.
5. The upper roof has a mild slope to drain off the rain water. It has one or two doors or windows which should only open either towards north or south to avoid the rays of sun falling directly on the thermometers.
6. This box is kept on a wooden stand at a height of about 4 feet above the ground surface.
7. The box and the stand are painted white throughout.
8. The Stevenson Screen is meant for keeping the instruments like minimum and maximum thermometers, self recording instruments namely, thermograph, barograph, hygrograph, etc.
9. This is available in two different sizes and the inside dimensions are given in table 6.1

TABLE 6.1

DIMENSIONS OF SINGLE STEVENSON SCREEN (SSS) AND DOUBLE STEVENSON SCREEN (DSS)

S. No.	Instrument	Length (cm)	Width (cm)	Panes height (cm)
1.	S.S.S.	45.5	27	38
2.	D.S.S.	91.0	27	38

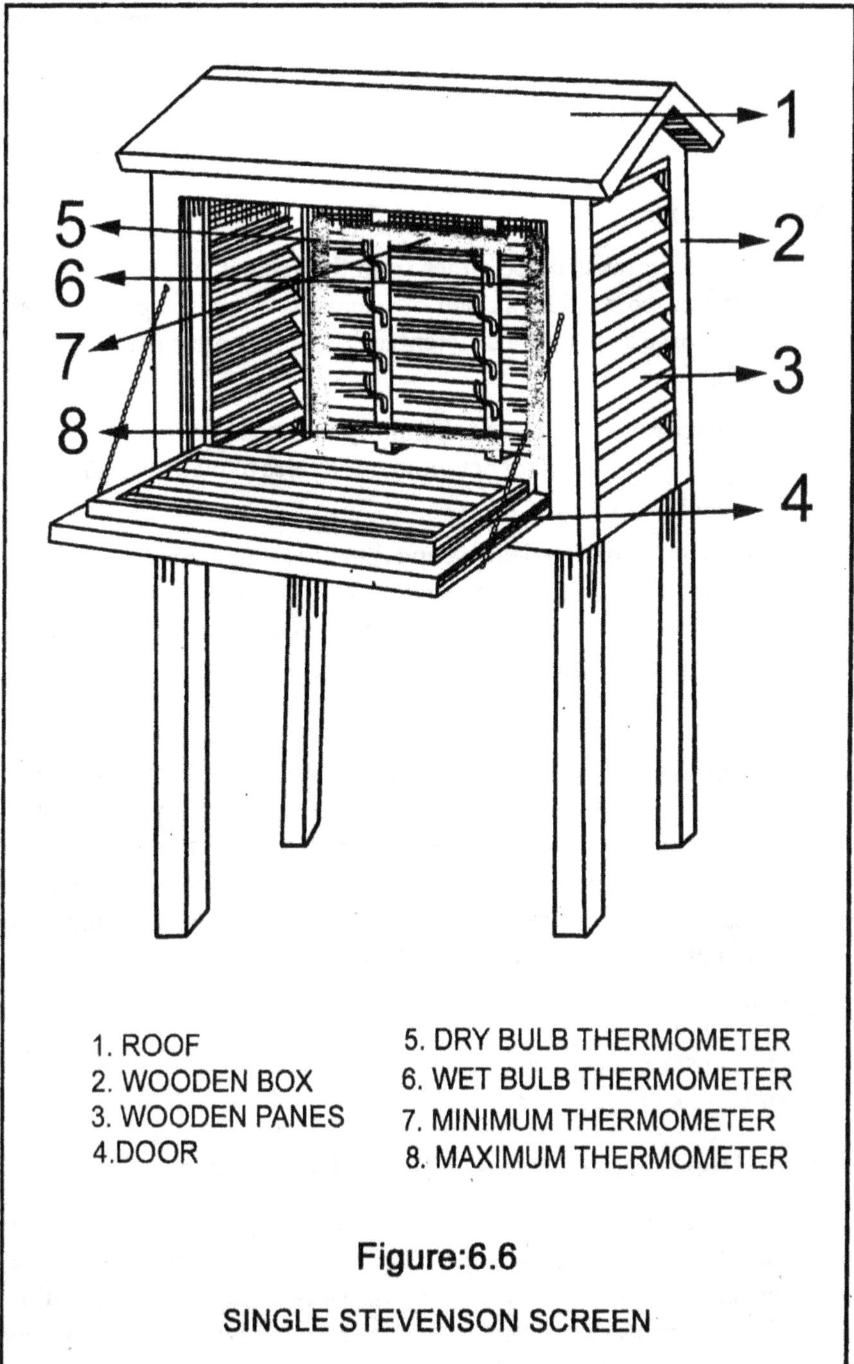

Figure:6.6

SINGLE STEVENSON SCREEN

Minimum and maximum thermometers are kept horizontally where as the wet and the dry bulb thermometers are kept vertically on either side of minimum and maximum thermometers.

Basic Humidity Terminology

Water Vapour

1. Evaporation takes place either from the ground surface or from its vegetation cover. The stream of vapour is directed upward. Water returns to earth in quite a different fashion, being precipitated in liquid or solid form.
2. The water vapour content or humidity, is a measure of the dryness of the air. It is an important determinant of rates of evaporation (E) and transpiration (T).
3. Stomata of some species respond directly to the humidity of the air, independently of the leaf water status. Thus, stomata close in dry air restricting exchange of carbon-dioxide and water vapour, and thus possibly reducing growth.

Saturation

It is defined as, "The condition of atmosphere at which the maximum amount of water vapour is held at a particular temperature". When air is completely saturated then there will be no evaporation. Under the saturated conditions, the dry and the wet bulb thermometers indicate same temperature. The saturation capacity of air increases with increase in temperature and vice-versa.

Soil Moisture

The moisture contained in the soil above the water table including the water vapour present in the soil pores.

Results of Research on Effect of Humidity on Soybean Crop

1. Yield of soybean as obtained from adequate supplies of soil moisture are affected by the atmospheric humidity. Reduction in yield by 21 per cent recorded for soybean grown at day/night relative humidities of 47/46 per cent as compared to 81/84 per cent due to flower abortion at decreased humidities during reproductive stage.

2. At low atmospheric humidity reduced plant growth rate was recorded in tropics as a consequence resulted in low rates of photosynthesis, even under irrigated conditions.
3. Foliar application of any chemicals is ineffective under high humid conditions (more than 60 per cent) in controlling leaf eating larvae on soybean.

Units of Measurement

Already given against each term.

Chapter - 7

Evaporation and Transpiration

"Such a person must factually know the greatest of all, the Personality of God-head, who is unembodied, omniscient, beyond reproach, without veins, pure and uncontaminated, the self-sufficient philosopher who has been fulfilling everyone's desire since time immemorial".

EVAPORATION

The evaporation is defined as, "A physical process in which liquid water is converted into its vapour" The sun is the source of energy that activates the hydrologic cycle i.e., the heat required for evaporation is supplied by the sun. The moisture in the atmosphere is supplied by evaporation. In this process the molecules of water having sufficient kinetic energy to overcome the attractive forces tending to hold them within the body of liquid water are projected through the water surfaces.

Importance of Evaporation Crop Plants

1. Evaporation is an important process of hydrologic cycle.
2. The evaporation from the soil is an important factor in deciding the irrigation water requirements of a crop.
3. In modifying the micro-climate of a crop the evaporation from the soil is an important factor for consideration.
4. Evaporation is the most important of all the factors in the heat budget, after radiation.
5. The evaporation is also one of the most important factors in the water economy.
6. Since, a certain amount of evaporation also demands a definite amount of heat, it provides a link between water budget and heat budget.

Factors Affecting the Evaporation

The evaporation losses from a fully exposed water surface are essentially the functions of several factors.

I. Environmental Factors

1. Water Temperature

With an increase of temperature the kinetic energy of water molecules increases and surface tension decreases. So, the rate of the evaporation increases with a rise in the temperature. The maximum amount of water vapour that can exist in any given space is a function of the temperature.

2. Wind

The evaporation from a fully exposed surface is directly propostional to the velocity of wind and *vice-versa*, because the dry wind replaces the moist air near the water. The process of evaporation takes place continuously when there is a supply of energy to provide latent heat of evaporation (approximately 540 calories per gram of water evaporate at 100°C).

3. Relative Humidity

A mechanism exist to remove the vapour above the surface so that the vapour pressure of the water vapour in the moist layer adjacent to the liquid surface is less than the saturated vapour pressure of the liquid (vertical gradient of vapour pressure). When the air above water is dry or has low relative humidity, the evaporation will be greater than when air has high relative humidity over the water.

4. ***Pressure***

The evaporation is more at low pressure and *vice-versa*.

II. Water Factors

1. ***Composition of Water***

The dissolved salts and other impurities decreases the rate of the evaporation. The evaporation is inversely proportional to the salinity of water. The rate of evaporation from the surface of the sea is less than that of fresh water in rivers. Under equivalent conditions ocean water evaporates 5 per cent less than fresh water in rivers.

2. ***Area of Evaporation***

If two volumes of water are equal in two containers, the evaporation will be greater for the one having the larger exposed surface.

Measurement/Estimation of the Evaporation

The evaporation from a surface is influenced not only by environmental factors but also by the depth, size, state of the evaporating surface, surroundings, etc. The following methods are used to estimate the amount of evaporation from a free water surface.

1. Energy budget method.
2. Mass transfer method.
3. Water budget method.
4. Empheirical formulae.
5. Evaporation measurements using pans.

There are four main types of evaporimeters or pans used for measuring evaporation.

1. Floating Pans

(a) These are made to float in water bodies with special devices.

(b) The loss of water from these pans is equal to the loss of water from the water bodies in which they are floating.

(c) These are costly and under windy weather, the accuracy is reduced.

2. Pans Placed Above the Ground

(a) These are installed on the ground.

(b) They include, 20 square metre evaporimeter, G.G.I. 3000 the widely used U.S. Weather Bureau Class 'A' pan evaporimeter, etc.

(c) The major draw back in these instruments is the influence of sensible heat flux on the sides and bottom of the pans.

(d) Increased rates of evaporation is a common feature.

3. Sunken Pans

Inspite of problems like cleaning and heat leakage, these are most commonly used by micro-meteorologists in crop weather studies (further details are given in the later part of this chapter).

4. Lysimeters

(a) These are very expensive and can not be moved from one place to another in the same field laboratory.

(b) These are used to measure not only the evaporation but also the evapotranspiration.

The Most Commonly used Evaporimeters

1. U.S.W.B. Class 'A' open pan evaporimeter.
2. Sunken screen evaporimeter.

1. U.S.W.B. Class 'A' Open Pan Evaporimeter (Mesh covered fixed point gauge)

Principle

The amount of water lost by evaporation from the free water surface in the pan at any given interval of time is measured by adding known quantities of water to the pan and bringing it to the original level.

Operation and Measurement

This is an instrument used to measure the amount of water lost by evaporation per unit area at a given interval of time. The values of evaporation give a measure of evaporative power of the air layers near the ground (Figure 7.1).

1. This is made-up of galvonised iron or copper sheet of 20 gauge thickness. The 10 mm thick copper is the latest recommended standard.
2. The pan of the evaporimeter is 122 centimetres in diameter and 25.5 centimetres in deep.
3. It is painted white and is covered with a lid of hexagonal mesh to protect the water from birds and squirrels.
4. This is used to measure the rate of evaporation in mm/day with a precision of 0.1 mm.
5. Water level in this appliance should be maintained upto 20 centimetres.
6. In order to provide undisturbed water surface, a still-well is used. It is kept in the pan at the base and is provided with 3 small openings (120 degrees apart) at its bottom so that the water level of the pan corresponds to that of still-well.
7. The reference point is provided by the brass rod, fixed at the centre of the still-well and is tampered to end at a point exactly 190 mm above the base of the pan.
8. Measured quantities of water is either added or removed to bring back the level of water to its original position.
9. The rate of evaporation is determined by using the equation : Volume = Area × Depth.
10. A hook gauge is used for measuring the rate of evaporation. It works on the principle of screw gauge and the least count is 0.1 mm.
11. The pan rests on a wooden platform which is painted white and placed about 10 centimetres above the ground surface.
12. This allows free circulation of air and also to detect leakages, if any..
13. A thermometer to measure the temperature of the water is fixed with a clamp to the side of the pan so that the bulb dips 5 centimetres below the water surface.

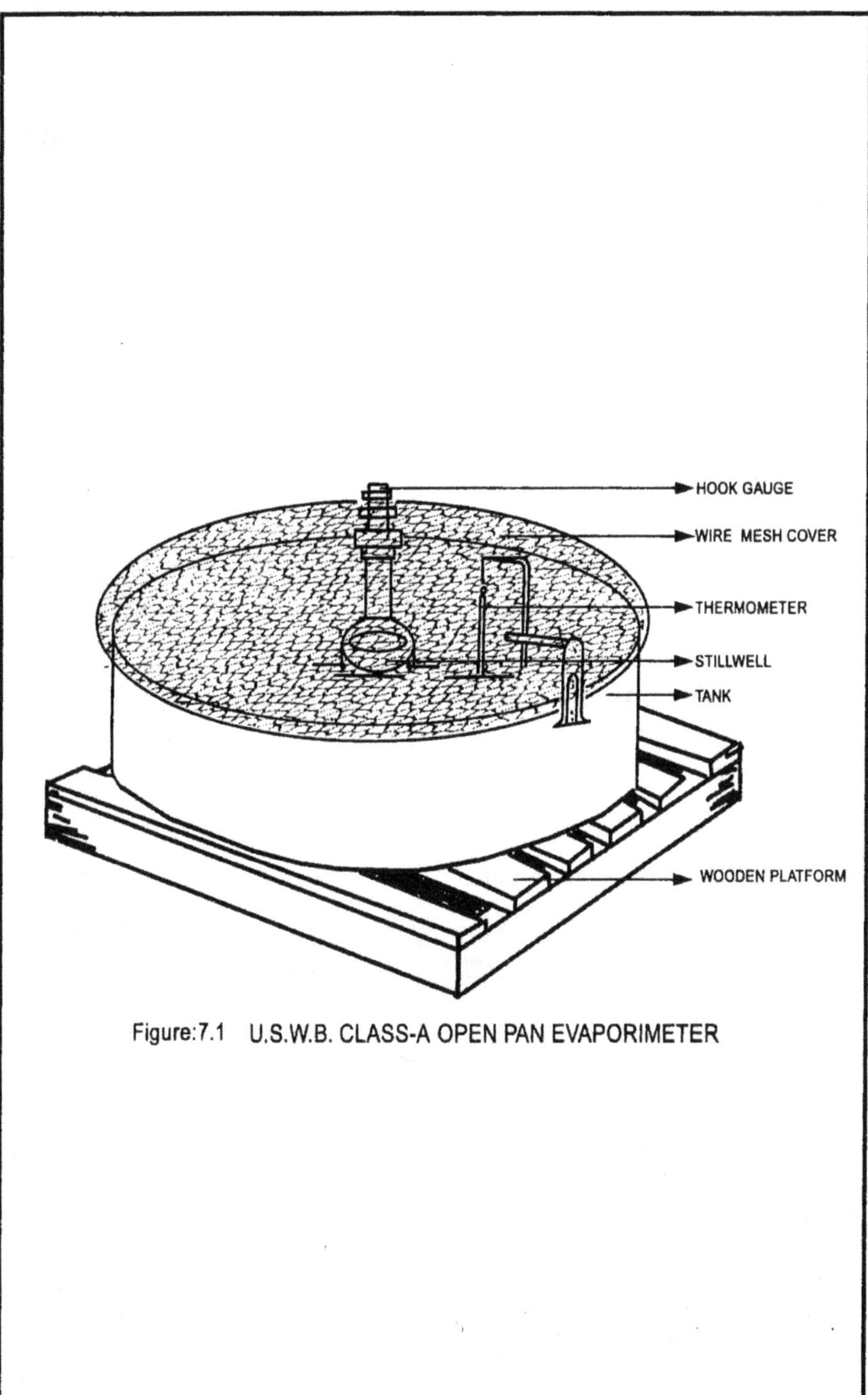

Figure:7.1 U.S.W.B. CLASS-A OPEN PAN EVAPORIMETER

14. If rain is there, the water level in the pan increases. So, water has to be removed to bring back its level to original position. After knowing the depth of water it is easy to find evaporation (if any) as the depth of the rainfall is already known from the rain gauge.

15. The measuring cylinder is a brass container with scale ranging from 0-20 cm.

16. The diameter is exactly one tenth of that of the pan i.e., 122 mm, which means the cross sectional area of the cylinder is exactly one hundredth that of the pan.

17. So, 200 mm of water from the cylinder added to the pan will raise the level of water in the pan by 2 mm.

18. The amount of water lost by evaporation from the pan divided by the time interval gives the rate of evaporation.

19. Since, the capacity of cylinder is only 20 cm, the cylinder has to be filled more than once if over2 mm of water is lost by evaporation.

20. Calibrated charts are also available to enable the semi- skilled workers to observe the corresponding evaporation. Observations with evaporimeter should be taken twice a day at 0830 and 1430 hours IST.

The observations are to be taken as detailed below

(a) Read the thermometer when just immersed in the water.

(b) When water level is below the reference point, add water to the evaporimeter using the measuring cylinder.

(c) Add water until the tip of the fixed point equals the surface of the water in the still-well.

(d) For example, one full cylinder and 10 cm i.e., 30 cm of the water is added to the pan, this divided by 100 i.e., 3.0 mm is the amount of water lost by evaporation from the pan, if no rainfall occured since the last observation.

(e) On a rainy day, if the amount of water taken out to bring the level equal to point is 38 cm, the difference as per the above calibration in the description is 3.8 mm. If the rainfall is 5.7 mm during the day, then the evaporation is 5.7 - 3.8 = 1.9 mm.

(f) When there is light rain, and the water level may not rise then the procedure to be followed is like this. If 20 cm of water is added to the pan (i.e., 2.0 mm water calibration) and rainfall is 1.2 mm, then the actual evaporation is 2.0+1.2 = 3.2 mm.

Precautions to be taken include

(a) Repairs for any leaks must be attended as and when noticed.

(b) Clean the pan and still-well regularly.

(c) Paint the evaporimeter with white paint every year.

(d) Use lemon juice to remove white deposits on the bulb of the thermometer.

(e) To avoid algal growth add few grams of copper sulphate.

2. Sunken Screen Evaporimeter

Principle

A still-well is attached to the evaporimeter through a connecting tube. The whole instrument is burried into the ground. Measured quantities of water is either added or removed to indicate the evaporation.

Operation and Measurement

1. This instrument comes under the category of sunken pans. This was developed by Dr. Sharma and Dastane in the year 1966 at Indian Agricultural Research Institute, Pusa, New Delhi. This is very useful in agrometeorological observatories and is more dependable than other instruments (Figure 7.2).
2. The pan of this evaporimeter is made up of galvonised iron sheet of 20 gauge thickness. It has a diameter of 60 centimetres and depth of 45 centimetres.
3. A still- well with 15 centimetres diameter and 45 centimetres depth is attached to the pan through a connecting tube.
4. This evaporimeter is painted white throughout.
5. Both the pan and still-well are covered with a lid of hexagonal mesh.
6. Water level in this evaporimeter is maintained upto 35 centimetres from the bottom of the pan.

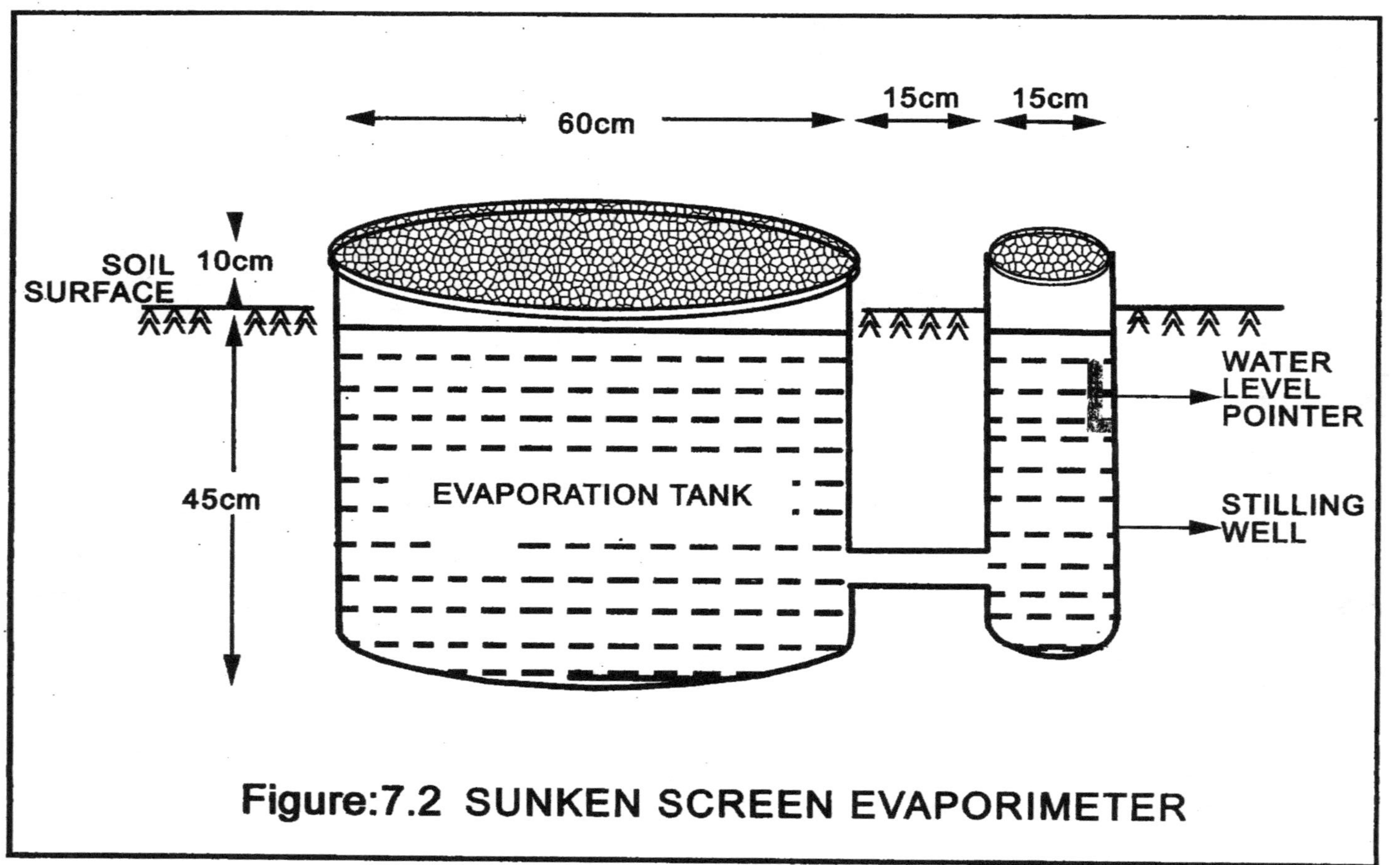

Figure:7.2 SUNKEN SCREEN EVAPORIMETER

7. A pointer attached to the wall of still-well and bent upwards at right angles to the well is used to maintain the water level.

8. The tip of the pointer is at a height of 35 centimetres from the bottom. The whole instrument is put into the ground upto a depth of 35 centimetres.

9. Measured quantities of water is either added or removed to bring back the level of water to its original position.

10. The rate of evaporation is determined by using the equation : Volume = Area x Depth.

11. Calibrated charts are also available for ready reference.

12. This evaporimeter is preferred over the U.S.W.B. class 'A' open pan evaporimeter, because the crop co-efficient range is very small i.e., 0.95 to 1.05.

13. With the help of this crop factor, water requirement of the crop can be worked out and scheduling of irrigation is done.

The differences between U.S.W.B. class 'A' open pan evaporimeter and sunken screen evaporimeter are given in Table 7.1.

TABLE 7.1

THE DIFFERENCES BETWEEN U.S.W.B. CLASS 'A' OPEN PAN EVAPORIMETER AND SUNKEN SCREEN EVAPORIMETER

S. No	Character	U.S.W.B. Class 'A' open pan evaporimeter	Sunken screen evaporimeter
1.	Place of instrument	Above the ground surface	Below the ground
2.	Diameter of the pan	122 cm	60 cm
3.	Height of the pan	25 cm	45 cm
4.	Place of stilling well	Inside the pan	Outside the pan
5.	Crop co-efficient factor	0.5 to 1.3	0.95 to 1.05
6.	Rate of evaporation measured	At free surface	In the cropped field

TRANSPIRATION

Transpiration is defined as, "The loss of water from living parts of the plant". There are 3 kinds of transpiration.

1. Stomatal Transpiration

This is the most common method of transpiration in all plants. In this kind of transpiration water evaporates through small openings called stomata. The stomata are present on green parts of the plant, mainly leaves.

2. Cuticular Transpiration

Outside the epidermal cells of a leaf, there is a thin layer called cuticle. The loss of water through cuticle is known as cuticular transpiration. The thickness of the cuticle and presence or absence of wax coating on the surface of the leaves affect the amount of water loss. Only 5 to 10 per cent of water transpires in this kind of transpiration.

3. Lenticular Transpiration

The openings present on the older stems are known as lenticels and the kind of water loss through these openings is known as lenticular transpiration. Nearly one per cent of the total loss of water in a plant takes place through this kind of transpiration.

Importance of Transpiration on Crop Plants

The following three roles are played by transpiration.

1. Dissipation of radiant energy by plant parts.
2. Translocation of water in the plants.
3. Translocation of minerals in the plant.

Factors Affecting the Transpiration

I. Environmental Factors

1. Light

Light plays predominant role in transpiration both directly and indirectly. The direct effect of light is on the opening and closing of stomata. Bright light is the main stimulus which causes stomata to open. It is because of this reason that all plants show a daily periodicity of transpiration rate. The indirect effect of light is that the increasing light intensity raises the temperature of leaf cells. This increases the rate at which liquid water is transformed into vapour.

2. ***Atmospheric Humidity***

The rate of transpiration is almost inversely proportional to atmospheric humidity. The rate of transpiration is greatly reduced when the atmosphere is very humid. However, as the air becomes dry, the rate of transpiration also increases proportionately. These effects occur in accordance with the law of simple diffusion.

3. ***Air Temperature***

Increase in the temperature results in opening of stomata. Temperature has significant effect on the permeability of the wall of the guard cells and therefore greatly effect the osmatic phenomenon. This phenomenon is responsible for the movement of guard cells.

4. ***Wind Velocity***

The velocity of the wind affects the rate of transpiration to a greater extent. Fast moving wind and air currents bring fresh and dry masses of air in contact with leaf surfaces. So, higher the wind speed higher the transpiration.

II. Plant Factors

Some plants adopt physiological modifications to check the excess transpiration. Some other plants modify their structure for this purpose, thereby withstand drought. Such characters greatly effect the transpiration.

1. ***Plant Height***

The water need of a crop varies with its height. In general, the rate of transpiration of a tall crop will be more (around 50 per cent) than when the crop is cut or clipped to half.

2. ***Leaf Characteristics***

In some plants like *Cacti* and other desert plants leaves are altogether absent and their functions are taken up by the stem itself. In case of *Pines*, *Firs,* etc., the leaf size is very much reduced. In such cases reduction in leaf area brings about reduction in transpiration. Some graminaceae family plants (maize), flower plants, etc., roll up or turn the edges of their leaves when exposed to bright sun and fast breeze. This causes reduction in the transpiration.

3. *Availability of Water to the Plant*

If there is little water in the soil, the tendency for dehydration of leaf causes stomatal closure and a consequent fall in transpiration. This situation occurs during :

(a) Periods of drought.

(b) When the soil is frozen.

(c) At a temperature so low that water is not absorbed by roots.

Measurement of Transpiration

Transpiration shall be measured in the laboratory where evaporation is eliminated and water losses are found by weighing. The most commonly used methods of measurement of transpiration are :

1. ***Weighing method*** **:** This method involves weighing fresh cut parts of the plant first, and later periodically until wilting starts.
2. ***Potometer method*** **:** This is an instrument to measure transpiration as accurately as possible in laboratories.

There are some notable differences between evaporation and transpiration (Table 7.2).

TABLE 7.2

DIFFERENCES BETWEEN EVAPORATION AND TRANSPIRATION

S. No.	Evaporation	Transpiration
1.	Controlled by meteorological factors.	Controlled by both meteorological and plant factors.
2.	Diffusive resistance is absent.	Diffusive resistance occurs due to internal leaf geometry and presence of stomata.
3.	Also occurs in night under advective heat transportation.	Reduced in the night due to closer of stomata.
4.	This is purely a physical phenomenon which takes place from any exposed surface with moisture.	This is a physiological phenomenon which takes place only in living plants.
5.	This takes place through any openings or pores.	This takes place through guard cells of stomata, cuticle, lenticules, etc.

Basic Evaporation Terminology

Evapotranspiration (ET)

1. Evapotranspiration denotes the quantity of water transpired by plants or retained in the plant tissue plus the moisture evaporated from the surface of the soil.

2. As long as the rate of root uptake of soil moisture balances the water lost from the canopy (moisture balances the water lost from the canopy by transpiration), evapotranspiration continues to occur at its potential rate.

3. When the rate of root water uptake falls below the transpiration demand, actual transpiration begins to fall below the potential rate.

4. This is either because the soil can not supply water to roots quickly or the plant can no longer extract water to meet the evaporational demand.

Reference Evapotranspiration (ETo)

This represents the maximum rate of evapotranspiration from an extended surface of 8 to 15 centimetres tall green grass cover, actively growing and completely shading the ground under limited supply of water. Among the different methods of calculating ETo, the well-known ones are Penman method, Blaney – Criddle method, Thornthwaite method, etc.

Potential Evapotranspiration (PET)

Potential evapotranspiration (PET) for any crop is obtained from reference evapotranspiration and crop factors (Kc) when water supply is unlimited.

$$PET = Kc \times ETo$$

Actual Evapotranspiration

1. When the soil moisture becomes limiting, the actual evapotarnspiration falls below the potential rate.

2. In such a case, plant water uptake under different conditions of available soil moisture can be considered to estimate actual evapotranspiration.

3. Soil water is not equally available throughout a definable range of wetness from an upper limit (field capacity) to lower limit (permanent wilting point) both of which are characteristics and constants for a given soil.

4. A critical point some where between the field capacity and wilting point occurs, and above this ETo occurs at its potential rate, below which it is a decreasing function of moisture content..

Measurement of Potential Evapotranspiration (PET)

The methods of measurement or determining or estimating potential evapotranspiration fall into five categories.

1. Direct measurements by lysimeter.
2. Empherical formulae using one or more common climatic factors.
3. The aerodynamic approach.
4. The energy budget approach.
5. The use of evaporimeters.

Eventhough all the above methods do estimate PET a few of these methods are primarily research tools for the better understanding of the physical process of transfer of water. The other methods can be used in the field as a guide to daily operation (operational).

Importance of Evapotranspiration and Potential Evapotranspiration for Crop Plants

The studies on evapotranspiration and potential evapotranspiration are useful in :

1. Estimation of the soil moisture there by planning irrigation schedule of crops.
2. Understanding relationship between the crop yield and irrigation water.
3. Guiding for the production of a crop with a fully developed canopy.

Units of Measurement

Evaporation is expressed in units of depth i.e., millimetres or centimetres per day.

Chapter - 8

Rainfall

"Only one who can learn the process of nescience and that of transcendental knowledge side by side can transcend the influence of repeated birth and death and enjoy the full blessings of immortality."

PRECIPITATION

For general use the terms precipitation and rainfall are used as synonyms with each other. Precipitation is defined as, "Earthward falling of water drops or ice particles that have formed by rapid condensation in the atmosphere and are too large to remain suspended in the atmosphere". In condensation the water vapour is suspended in the air in different forms. But, in the precipitation an appreciable deposit either in solid or liquid form takes place on the earth surface. There are some common forms and different types in precipitation (Table 8.1).

TABLE 8.1

DIFFERENT FORMS AND TYPES OF PRECIPITATION

S. No.	Form	Type
1.	Liquid	Rain, Drizzle and Shower
2.	Solid	Snow and Hail
3.	Mixed	Sleet and Glaze

I. Liquid Forms

1. Rain

It is defined as, "Precipitation of drops of liquid water". The cloud consists of minute droplets of water and when these droplets combine and form large drops and can not remain suspended in the air they fall down as rain. These droplets are formed by rapid condensation. The size of rain drop is more than 0.5 m.m. in diameter. The imaginary lines drawn on a map connecting the points of equal rainfall are known as the "Isohytes".

2. Drizzle

It is more or less uniform precipitation of very small and minute rain drops. These drops can be carried away even by light winds. The diameter of drizzle drop is less than 0.5 m.m. It falls from low lying nimbostratus cloud. Fog merges to form drizzle.

3. Shower

It is the precipitation lasting for a short time with relatively clear intervals.

II. Solid Forms

1. Snow

It is defined as, "Precipitation of water in solid form of small or large ice crystals". It occurs only when the condensing medium has a temperature well below freezing (0°C) temperature. It is also seen in the form of flakes which are aggregates of many crystals, formed due to sublimation of water vapour at sub-freezing temperatures.

2 Hail

It is a precipitation of solid ice. A strong convective column on a warm sunny day may cause the formation of pellets of spherical shape with concentric layers of ice, which is known as hail. Hail falls from cumulonimbus clouds and is often associated with thunder and storm. The size of hail ranges from peanut to cricket ball. The rainfall associated with the hail is called as the "Hail storm".

III. Mixed Forms

1. Sleet

It is the simultaneous precipitation of the mixture of rain and snow. Occasionally, half frozen drops also fall as sleet forms when rain drops are frozen as they fall through a layer of cold air.

2. Glaze

Freezing rain is known as glaze. This is formed at sub-freezing temperatures when rain falls on objects or on ground. It looks like a sheet or coat.

Mechanism or Process of Rain Formation

1. The drops formed in a cloud are not all of the same size due to differences in the rate of condensation taking place in different places of the clouds.
2. As the air moves above, the large and smaller drops do not follow the same path.
3. The smaller droplets overtake the larger drops and suffer collision resulting into coalescence or combination.
4. These resulting droplets then attain a bigger size and fall down with greater terminal velocity.
5. These larger drops, therefore, during their downward journey suffer many collisions and grow in larger sizes.
6. Very small droplets can not suffer collisions and do not grow by coalescence.
7. The larger drops fall faster and ascend slower than the smaller ones.
8. This process is known as collision and coalescence mechanism.

9. There is another mechanism called the "Bergeron mechanism" through which the process of rain formation can be explained.

Types of Rainfall

There are mainly three types of rainfall.

1. Convectional Rains

1. The air near the ground becomes hot and light due to heating. Then it starts upward movement. This process is known as convection. (This differs slightly from "Convection" defined in Chapter 2.).
2. As the air moves upward it cools at about 10°C per kilometre i.e., at dry adiabatic lapse rate.
3. As it becomes saturated, relative humidity reaches to 100 per cent and dew point is reached where the condensation begins. This level (height) is known as condensation level.
4. Above this level, air cools at about 4°C per kilometre which is slightly lesser than saturated adiabatic lapse rate.
5. First, cloud is formed. Then, the further condensation results into precipitation. These rains are known as convectional rains and mostly occurs in the tropics.

2. Orographic Rains

1. When moist air coming from the sea or ocean strikes mountain it can not move horizontally. It has to overcome the mountains.
2. When this air rises upward, it cools down, cloud is formed and condensation starts giving the precipitation.
3. These rains are known as orographic rains.
4. These are also known as the "Relief rains" as the rains also occurs when the air from sea or ocean strike or pass over relief barriers.
5. Due to these processes rains with high intensity are possible on the windward side of the mountain.

3. Cyclonic and Frontal Rains

1. The rains received from the cyclones are known as the "Cyclonic rains" (Chapter 6).
2. When two opposing air currents with different temperatures meet, vertical lifting takes place.
3. This convection gives rise to condensation and the precipitation which is known as the frontal precipitation.

The size of the rain drop reaching the ground depends upon the following points.

1. The effect of friction of air on the falling drop (more the friction lesser is the size of rain drop).
2. The amount of evaporation undergone by the rain drop in its descent.
3. Air temperature.

Importance of Rainfall (Water) on Crop Plants

One centimetre of rain over an area of one hectare or 100 m^3 (100,000 litres) contains 4,339 grams.of oxygen at 20°C. This is equivalent to 3,000 litres of pure oxygen at atmospheric pressure. Consequently, a rain usually has a much more invigorating effect on a crop than does an irrigation. Rain water has extraordinary qualities.

1. Water has high solvent power and this plays an important role in crop plants as the plants get their nourishment from soil only in solution form.
2. Water plays an important role in life processes of crop plants (in the exchange of gases).
3. The heat capacity of water is high and its high thermal stability helps in regulation of the temperature of crop plants.
4. Water has highest heat conduction capacity and due to this the heat produced by the activity of a cell is conducted immediately by water and distributed evenly to all plant parts.
5. The viscosity of water is higher than that of many solvents and this property helps in protecting the crop plants and trees against mechanical disturbances.

6. Water is densiest at 4°C. The freezing point of fresh water is 0°C and that of sea water about – 2.5°C. So, the ice can float on the surface and plant life in deeper parts of sea is made possible.
7. The transparency of water facilitates the passage of light to great depths and this helps for the survival of aquatic plants.
8. The high surface tension that water has, helps in movement of water into and through the plant parts.
9. Rainfall influences the distribution of crop plants in particular and vegetation in general, as the nature of vegetation of a particular place depends on the amount of rainfall (the vegetation of a desert where rainfall is less differs a lot from the vegetation of a rainforest).

Action of Rain

(a) The rain, especially in hot countries like parts of Africa (where there is a heavy rainfall) has a powerful action in loosening and carrying away the soil. When the soil is bound together by roots of trees or grasses, loosening of soil is difficult but when trees are cut down and hill sides are cultivated, large areas are often completely washed away by heavy rainfall.

(b) Rain water absorbs a considerable proportion of carbon-dioxide to form carbonic acid. While passing through soil, carbonic acid can dissolve hard rocks such as those of limestone. Much of the water sinks into the ground and dissolves lime stone rocks forming underground caves.

Measurement of Rain

The rain can be measured with several instruments.

I. Symon's Rain Gauge

It is made up of galvanised iron sheet of 12 gauge thickness. Of late, fibre glass and plastic makes are also in use. This consists of four parts (a) Funnel (b) Body (c) Receiver and (d) Base.

Principle

The rain water entering the gauge from the top of the rim of the funnel is lead *via* funnel to the receiver. The rain water thus collected is measured with the help of a measuring cylinder.

Operation and Measurement

1. The diameter of the funnel is 12.7 c.m. The outer peripheral ring is made up of copper or brass and it is called as "Rim". It is designed in such a way that the rain water does not splashout (Figure 8.1).
2. The rain water received by the funnel is emptied into a collecting jar which is kept in an outer jacket or receiver.
3. The outer jacket is a cylindrical vessel closed at one end.
4. Besides housing the collecting jar, the outer jacket also receives the over flow of the rain water from it.
5. The funnel, the collecting jar, and the outer jacket are fitted into a base which has a locking arrangement.
6. The amount of the rain water is measured with the help of a calibrated glass measuring jar, corrected upto 0.1 m.m.
7. The rain guage should be kept on a hard compact levelled platform partially inserted in the ground in such a way that the rim is at a height of 1 foot (30 cm) above the ground surface.
8. The rim should be positioned on a perfectly horizontal plane. This can be done by using spirit level and rain gauge should be painted grey throughout.
9. Rain gauge should be checked for leaks and dust particles. Leaves should be removed from the receiver.
10. The measuring cylinder should be kept clean and a spare measuring cylinder should be available in the observatory.

II. Natural Syphoning Self Recording Rain Gauge

This is designed to give a continuous recording of the rainfall. This instrument not only records the total amount of rainfall that has fallen since the record was started but also the rate of rainfall. This is also known as pluviograph.

Principle

The rain water entering the guage from the top of the cover is lead *via* a funnel to the receiver consisting of a float chamber and a syphon chamber. A pen is mounted on the stem of the float and as the level of water rises in the receiver, the float rises and the

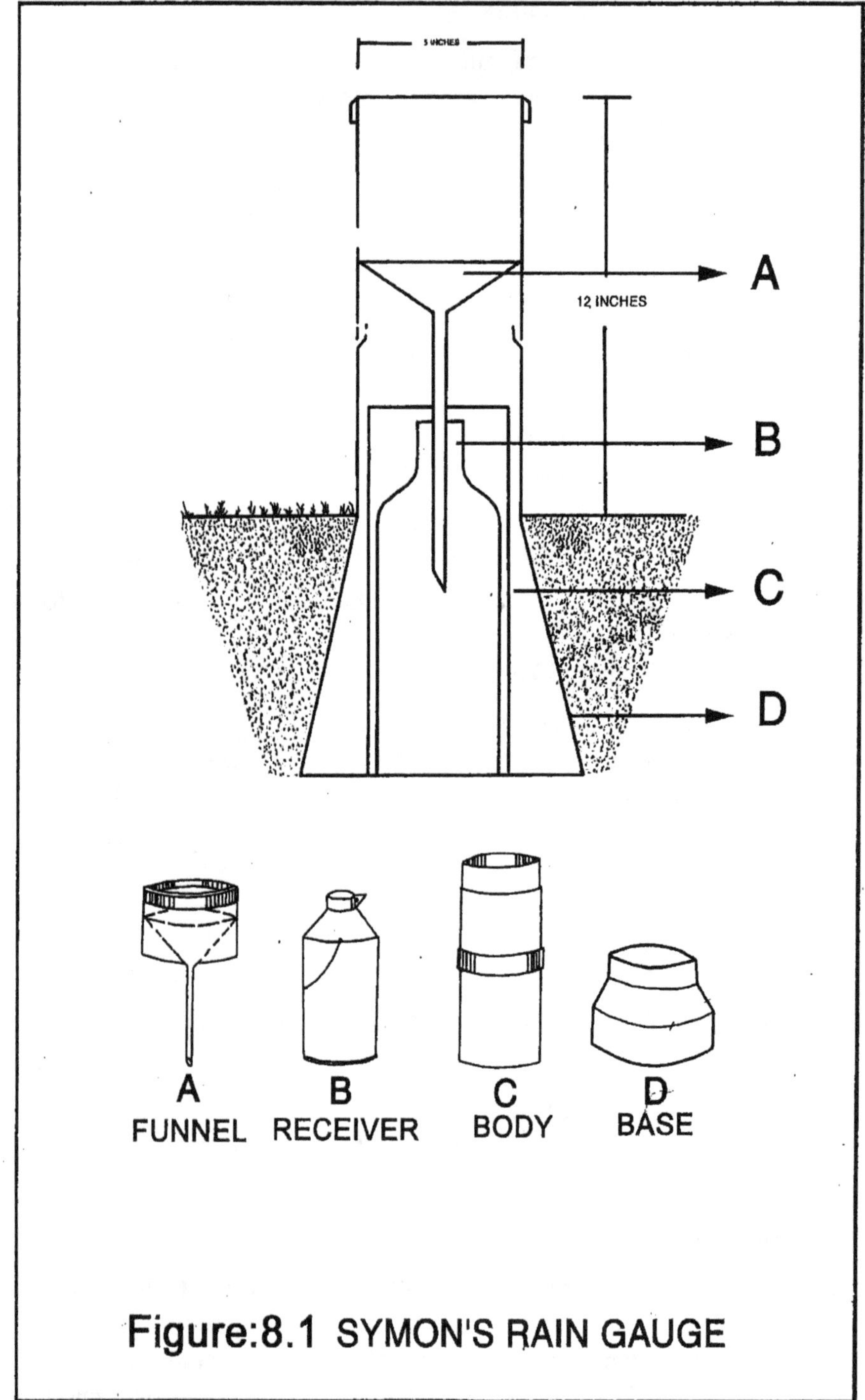

Figure:8.1 SYMON'S RAIN GAUGE

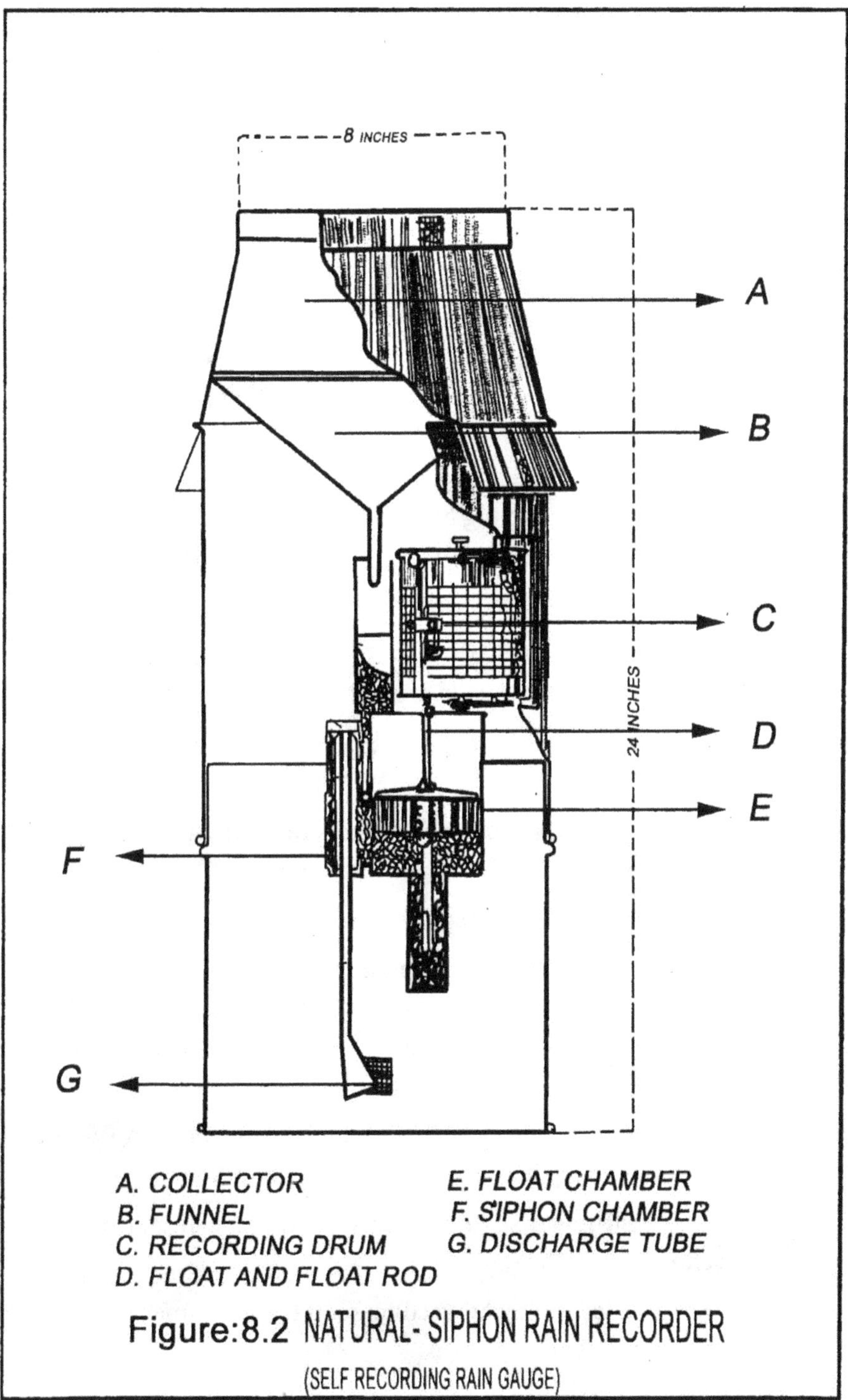

Figure:8.2 NATURAL- SIPHON RAIN RECORDER

(SELF RECORDING RAIN GAUGE)

pen records the level of the water in the chamber. Syphoning occurs automatically when the pen reaches the top of the chart.

Operation and Measurement

1. This is also made up of galvanised iron sheet of 12 guage thickness. Now-a-days fibre glass reinforced polyester material is extensively used.
2. This instrument also has a funnel with a glass rim of 203 millimetres. This acts as a lid and is provided with a slit or window. The rim of the funnel should be horizontal to the ground and exactly at a height of 75 centimetres above the ground (Figure 8.2).
3. The rain water received by the funnel is poured into a chamber through a connecting tube.
4. The chamber has a float and it is connected to a pen arm through a lever mechanism.
5. The tip of the pen arm is self-inked and touches a calibrated chart which is wrapped around a rotating drum.
6. The drum works with a clock mechanism and completes one rotation in 24 hours.
7. The X-axis of the chart represents rainfall with a precision of 0.1 m.m. and Y- axis represents time.
8. The chart has to be replaced only at the end of a rainy day.
9. This instrument has a syphoning mechanism and when the water reaches the maximum level it gets emptied automatically. The pen arm comes down to zero and rises again if there is further rainfall.
10. The chart has to be changed every day between 0830 and 0900 IST and there should be sufficient ink in the pen.
11. The instrument should always be kept clean and no leaves should enter the funnel.
12. A spirit level may be used to keep the instrument exactly as detailed above.

13. The slope of the trace of the pen at any point gives approximately the intensity of rain.
14. The rate of rainfall is obtained by dividing the total amount of rainfall with the total hours of the rainfall during a day.

III. Tipping Bucket Rain Guage

These rain gauges are generally attached to the automatic weather stations.

1. This is an automated device and accumulates a certain volume of water in a reservoir or bucket (Figure 8.3).
2. Normally, the reservoir or bucket is so small that it houses only 5 millimetres of rain water.
3. When the exact volume has been collected, the bucket automatically tips and the water is emptied.
4. This tipping mechanism also provides a method of recording on a moving chart or a counter attached to the tape recorder in the weather station.
5. Sensitivities upto 0.10 or 0.25 m.m. of rain are available.
6. The rim of rain gauge collector should be sharp and durable.
7. The main source of error is the exposure of the guage.
8. Depending upon the maximum intensity of rainfall for a given location, the tipping bucket rain Gauge with suitable sensitivity has to be selected.

CONDENSATION

1. Condensation is defined as, "The process in which the water vapour is converted into its liquid".
2. This process is inverse of the evaporation process. In condensation 600 calories of heat is released by each gram of water which was taken in the evaporation process.
3. Thus, the evaporation of water produces cooling effect and condensation gives warming effect.
4. The visible forms of condensed moisture in the atmosphere are known as "Hydrometeors"

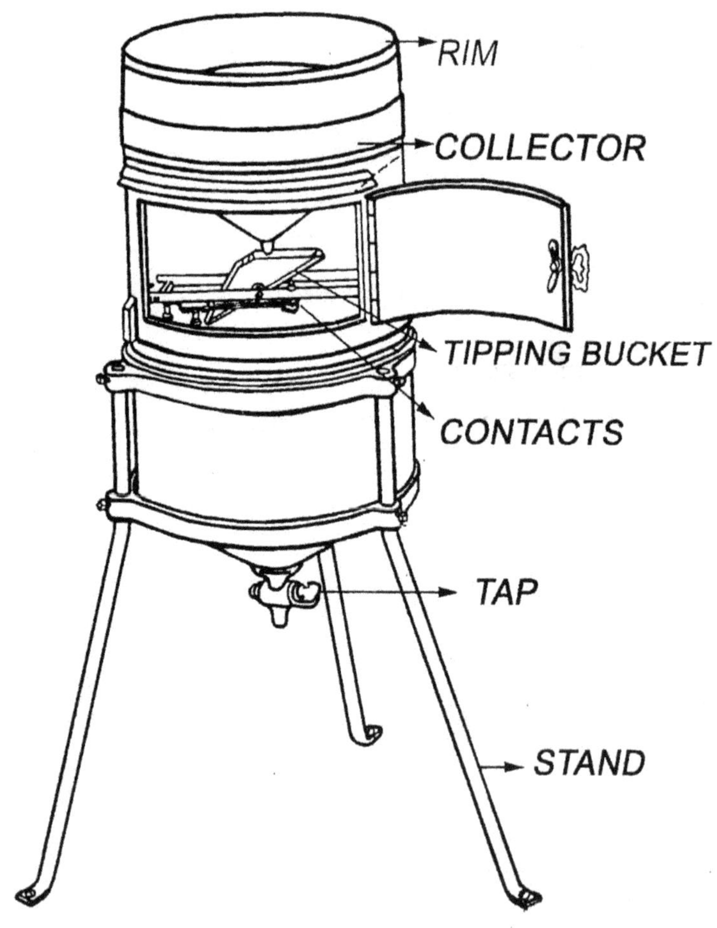

Figure:8.3

TIPPING BUCKET RAIN GAUGE

Conditions for Condensation

The following three conditions must be fulfilled for occurrence of condensation in the atmosphere.

I. Presence of Sufficient Water Vapour

- An adequate amount of water vapour is necessary to bring about saturation of air.
- Dew point shall be reached through this water vapour to begin condensation.

II. Presence of Condensation Nuclei

- Sodium chloride injected into the atmosphere by sea-spray, sulphur dioxide, nitrous oxide, etc., released from industries as combustion products and dust present in the atmosphere acts as nuclei of condensation.
- Water vapour can only deposit and condense on them as these are hygroscopic in nature (affinity to water).
- As these particles are microscopic or sub-microscopic in size (0.1 to 1 microns) these are called either hygroscopic nuclei or condensation nuclei.
- In the absence of hygroscopic nuclei, condensation can not trigger even if air is supersaturated and its temperature being below freezing level.

III. Cooling of Air

Cooling of air upto and below dew point is necessary for saturation of atmospheric air with water vapour.

Forms of Condensation

1 Dew

(a) The deposition of water vapour in the form of tiny droplets on the colder bodies by condensation is known as dew.

(b) The temperature at which water vapour condenses is known as dew point temperature.

(c) When the objects on the surface of the earth gets cooled in the night below dew temperature the water vapour is condensed on these surfaces.

(d) Dew forms when condensation takes place above freezing point.

(e) These surfaces should be good radiators and bad conductors of heat (Plant leaves, window glasses and pieces of paper).

The conditions for the formation of dew are :

- Clear skys
- Absence of wind

With particularly favourable conditions, dew deposition may commence before the sunset and continued after the sunrise. Deposition tends to be reduced under very calm conditions (wind speed range of 1-3 m/sec). Dew is an important secondary source of moisture for crops during the non-rainy season and plays vital role in plant growth. Dew occurrences benefit the plants in many ways.

Significance of Dew

1. The dew deposited on the leaf surface in the morning delays the rise in leaf temperature there by reduces the rate of evapotranspiration.
2. Dew provides water for direct plant use. The amount of dew deposition varies from 0.25 to 0.40 m.m. per night in semi-arid tropics (Usually, from September to April dew amounts are measured in these areas).
3. Under suitable conditions, in semi-arid areas, it may exceed even 25-30 m.m. per annum.

2. Frost

When the temperature of atmospheric air falls below 0°C before the dew point is reached, the water vapour is directly converted into crystals of ice called as "Frost". This is a form of sublimation, because, water vapour is directly converted into ice. Frost is injurious to agricultural and horticultural crops. There are two types of frost.

(a) Hoar frost

The white and opaque deposition in the form of ice crystals having the shapes of feathers, needles, etc., is known as hoar frost.

(b) Glazed frost

This is caused by rains which are freezed on falling to the ground. This is transparent.

3. Fog

"The extremely small water droplets suspended in the atmosphere reducing the horizontal visibility is known as "Fog". Fog reduces the visibility. The conditions for the formation of fog are :

(a) Calm wind.

(b) The relative humidity atleast upto 75 per cent.

Fog is also called as, "Cloud on and near the ground". There is no particular form, shape or structure to fog. The following types of fog are most prevalent in crop grown regions.

Radiation Fog

This results from rapid loss of night time radiation either from the ground or lower air. This radiation produces cooling.

Inversion Fog

This is another type of radiation fog. This results from condensation of water vapour in a mass of warm moist air lying over a layer of cold air near the ground.

These two fogs occur during night or cold morning. They disappear due to desaturation of air with vapours after the rise of sun in the morning.

Advection Fog

This fog occurs when warm moist air rises in over a cold surface of either land or water. These fogs occur at any time of the day.

4. Smog

The combined effect of smoke and fog droplets which reduces visibility is called "Smog". Some solid particles like dust, smoke from fires and industries restrict the visibility further when these are added to smog which is known as 'Haze'. On some occassions toxic materials present in fog, smog and haze which are harmful. All these processes cause difficulty in rail, road, aviation and shipping traffics.

5. Rime

This is "Freezing fog" and is formed when wet fog having super cooled droplets immediately freeze on striking objects having temperatures below freezing point. White ice is formed on windward freezing point (telegraph post).

6. Mist

Mist is less dense fog. The suspended water droplets in the atmosphere restricts the visibility between 1100 to 2200 yards or number 4 on the coded scale (IMD). The obscurity is known as mist. The relative humidity is 75 per cent when mist occurs. Mist disappears with rising sun.

Measurement of Condensation

Dew by Dew Gauge

1. The dew gauge consists of a rectangular block of wood 32 × 5 × 2.5 cm in dimensions placed at heights 5, 25, 50 and 100 cm above ground over iron supporters coated with a red paint which favours the retention of dew deposited on it (Figure 8.4).
2. This is painted with red-oxide.
3. The dew gauge is exposed at about sunset and the size, form and distribution of the dew deposited on the gauge is observed at about sunrise.
4. The appearance is comapred with a set of photographs of each dew type.
5. These photographs bear the dew scales 1 to 8, each number bearing its equivalent in mm of dew.
6. A direct method of measuring dew is to expose a weighed plate of hygroscopic material at sunset and re-weight it after sunriese.
7. The hydroscopic material may be gypsum, blotting paper, etc. This method requires accurate weighing and protection at sunrise to prevent evaporation.

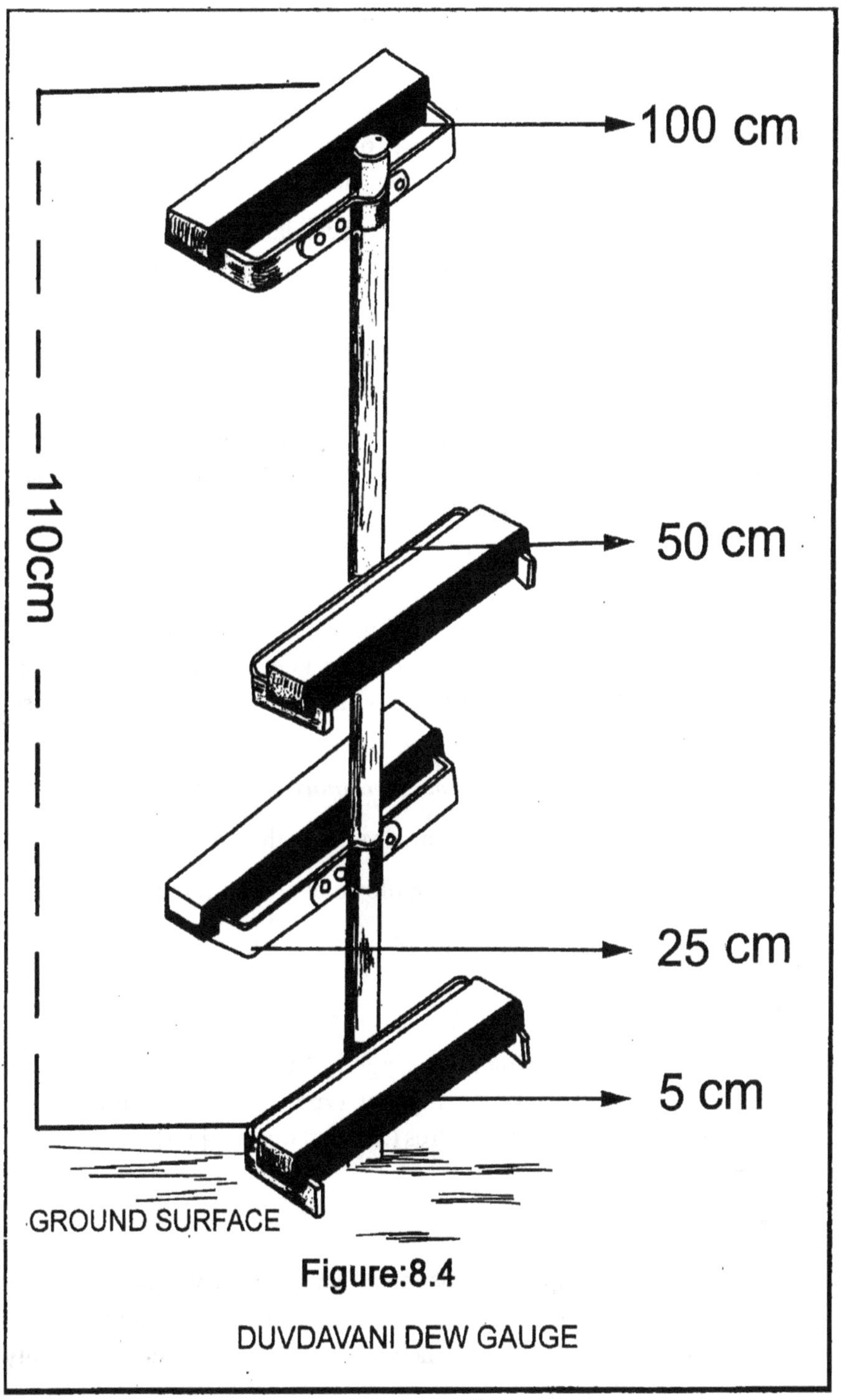

Figure:8.4

DUVDAVANI DEW GAUGE

8. Exposing filter paper " Sensor " with dew spots helps in qualitative assessment of dew.
9. When wetted by dew, the spots will sperad to an extent which depends on both the duration and intensity of dew fall. Now-a-days, dew duration recorders are also available.
10. Dew balancse are used to assess dew quantitatively using blotting papers. In this method the leaf surface also has to be measured.

Clouds

Definition

Cloud is defined as, "An aggregation of minute drops of water suspended in the air at higher altitudes". The rising air currents tend to keep the clouds from falling to the ground.

Cloud Formation

When air rises due to increase in temperature the pressure being less it expands and cools until temperature is equalised. If the cooling proceeds further till the saturation point, the water vapour condenses and cloud is formed.

Clouds are also formed under two more conditions

(a) When a current of warm air strikes the one that is cooler.

(b) When moist air from sea blows over a cold land.

Basic Types of Clouds

Cirrus (Ci)

Cirrus means 'curl' which is recognised by its veil like fibrous or feathery form. This is the highest type of cloud, ranging from approximately 7 to 12 kilometres (20-35 thousand feet) in altitude, in tropics and sub-tropics.

Cumulus (Cu)

Cumulus means heap or globular mass. This cloud is wooly and bunchy with rounded top and flat base. This is seen in summer months as it is formed due to convection. The height varies depending upon humidity of the atmospheric air.

Stratus (St)

This cloud looks like a sheet. This is the lowest in height from the ground.

Nimbus (Nb)

This looks dark and ragged. Precipitation occurs from this cloud. The prefix "Nimbus" means associated with precipitation and "Alto" means above normal height. Combination of different primary clouds are referred with these clouds.

WMO Cloud Classification

The World Meteorological Organisation (WMO) classified the clouds according to their height and appearance into 10 categories. From the height, clouds are grouped into 4 categories (*viz.*, family A, B, C and D) as stated below and there are sub-categories in each of these main categories (Figure 8.5).

Family A

The clouds in this category are high. The mean lower level is 7 kilometres and the mean upper level is 12 kilometres in tropics and sub-tropics. In this family there are three sub-categories.

1. ***Cirrus (Ci)***

- In these clouds ice crystals are present.
- Looks like wispy and feathery. Delicate, desist, white fibrous, and silky appearance.
- Sun rays pass through these clouds and sunshines without shadow.
- Does not produce precipitation

2. ***Cirrocumulus (Cc)***

- Like cirrus clouds ice crystals are present in these clouds also.
- Looks like rippled sand or waves of the sea shore.
- White globular masses, transparent with no shading effect.
- Meckerel sky.

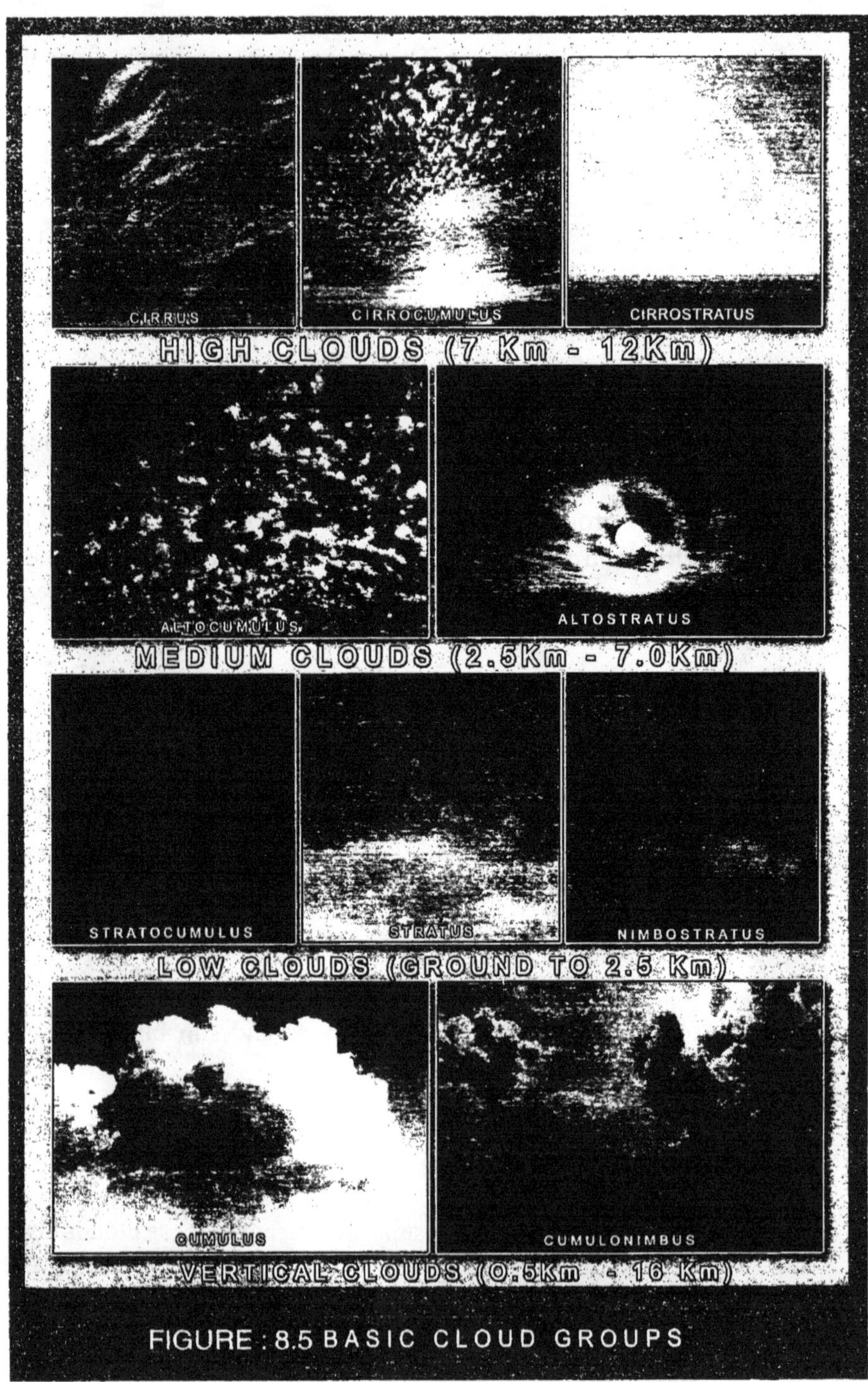

FIGURE : 8.5 BASIC CLOUD GROUPS

3. ***Cirrostratus (Cs)***

- Like the above two clouds ice crystals are present in these clouds also.
- Looks like whitish veil and covers the entire sky with milky white appearance.
- Produces "Halo".

Family 'B'

The clouds in this category are middle clouds. The mean lower level is 2.5 kilometres and the mean upper level is 7 kilometres in tropics and sub-tropics. In this family there are 2 sub-categories as detailed below:

1. ***Altocumulus (Ac)***

- In these clouds ice water is present.
- Greyish or bluish globular masses.
- Looks like sheep back and also known as flock clouds or wool packed clouds.

2. ***Alto-Stratus (As)***

- In these clouds water and ice are present separately.
- Looks like fibrous veil or sheet and grey or bluish in colour.
- Produces coronos and cast shadows.
- Rain occurs in middle and high latitudes.

Family 'C'

The clouds in this category are lower clouds. The height of these clouds extends from ground to upper level of 2.5 kilometres in tropics and sub-tropics. In this family, like high clouds, there are 3 sub-categorises.

1. ***Strato cumulus (Sc)***

- These clouds are composed of water.
- Looks soft and grey, large globular masses and darker than altocumulus.

- Long parallel rolls pushed together or broken masses.
- The air is smooth above these clouds but strong updrafts occurs below.

2. ***Stratus (St)***

- These clouds are also composed of water.
- Looks like fog as these clouds resemble greyish white sheet covering the entire portion of the sky (cloud near the ground).
- Mainly seen in winter season and occasional drizzle occurs.

3. ***Nimbostratus (Ns)***

- These clouds are composed of water or ice crystals.
- Looks thick dark, grey and uniform layer which reduces the day light effectively.
- Gives steady precipitation.
- Sometimes tooks like irregular, broken and shapeless sheet like.

Family 'D'

These clouds form due to vertical development i.e., due to convection. The mean low level is 0.5 and mean upper level goes upto 16 kilometres.

In this family two sub-categories are present.

1. ***Cumulus (Cu)***

- These clouds are composed of water with white majestic appearance with flat base.
- Irregualr dome shaped and looks like cauliflower with wool pack and dark appearance below due to shadow.
- These clouds usually develop into cumulo-nimbus clouds with flat base.

2. ***Cumulonimbus (Cb)***

- The upper levels of these clouds possess ice and water is present at the lower levels.
- These clouds have thunder head with towering envil top and develop vertically.

- These clouds produces violent winds, thunder storms, hails and lightening, during summer.

Cloud Observation

The accuracy of cloud observation depends on the experience of agrometeorologist. He observes clouds regularly because of the obvious relation of clouds in the hydrological cycle. Cloud cover is obtained by viewing the fraction of the sky covered. The terms used to express the sky cover are :

(a) Clear : No clouds or less then 1/10 of sky obscurred by clouds.

(b) Scattered : 1/10 to less than 6/10 cover.

(c) Broken : 6/10 to 9/10 cover.

(d) Overcast : more than 9/10 cover.

The amount of cloud is denoted by "Okta".

MONSOONS

1. The term monsoon is derived from an Arabic word "Mausim" means "Season".
2. There are different concepts to explain Indian monsoons. Of them the "Thermal concept" proposed by Halley is of more practical relevance than other concepts like aerological, Flohins, etc.
3. The two types of distinguished monsoons over India are :
   1. South-West monsoon (SW).
   2. North-East monsoon (NE)

1. South-West (SW) Monsoon

1. In summer the land mass of India heats quickly and develops a strong low pressure centre, particularly over north-west India during April and it exists upto September.
2. As the pressure over the adjacent oceans is high, a sea to land pressure gradient is established (Figure 8.6).
3. Therefore, the surface air flow is from the high pressure areas over the oceans towards the low pressure areas over the heated land.

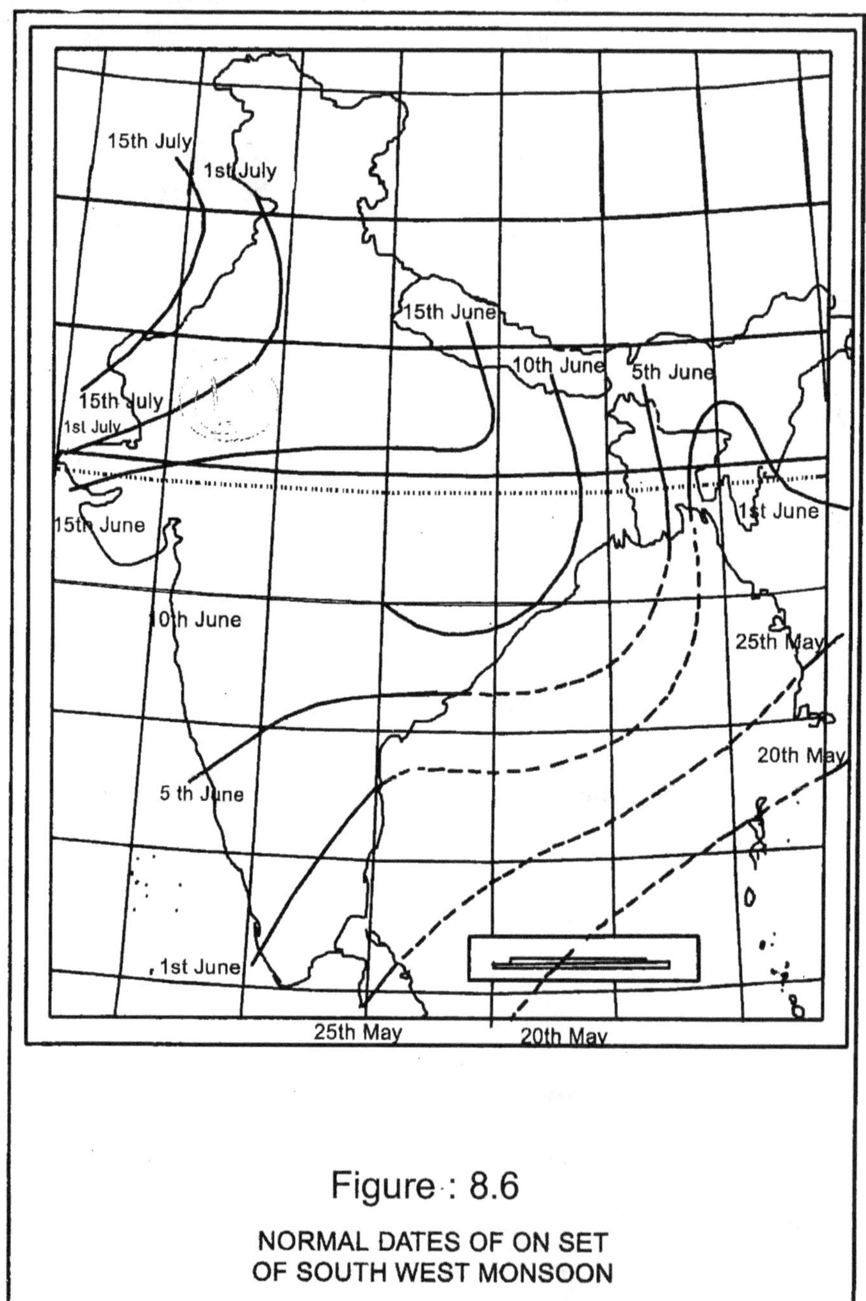

Figure : 8.6

NORMAL DATES OF ON SET
OF SOUTH WEST MONSOON

4. Eventhough, India should have north east monsoon winds throughout the year due to its position in NE trade wind zone the SW winds predominate because of existance of low pressure along Ganges and upper India.
5. The air that is attracted into the centres of low pressure from over the oceans is "Warm and moist".
6. This monsoon is active from June to September.
7. The rainfall received is 80 to 90 per cent of the total annual rainfall of India covering all parts.
8. This monsoon enters Kerala on June 1st and by 15th July it reaches the northern most parts of the country.
9. There are two branches of the South-West Monsoon.
   (a) The Arabian Sea branch : This branch crosses Western ghats.
   (b) The Bay of Bengal branch : This branch crosses Gangetic plains.

North-East (NE) Monsoon

1. A complete reversal of the south west monsoon winds takes place during winter (November - February).
2. In this season the land mass over India cools more rapidly than the surrounding oceans.
3. So, a strong high pressure centre develops over the continent (Figure 8.7).
4. On the other hand, the pressure over the adjacent oceans is relatively lower.
5. As a consequence, the pressure gradient is directed from land to sea and winds flow in North-East direction.
6. Therefore, there is an outflow of air from the continental land mass to the adjacent oceans.
7. The air flow brings "Cold dry" air towards low latitudes.
8. This monsoon is active from October to mid-December .
9. The rainfall received is 10 to 20 per cent of the total annual rainfall of India, covering parts of Andhra Pradesh (Nellore, Chittore) and Tamil Nadu.

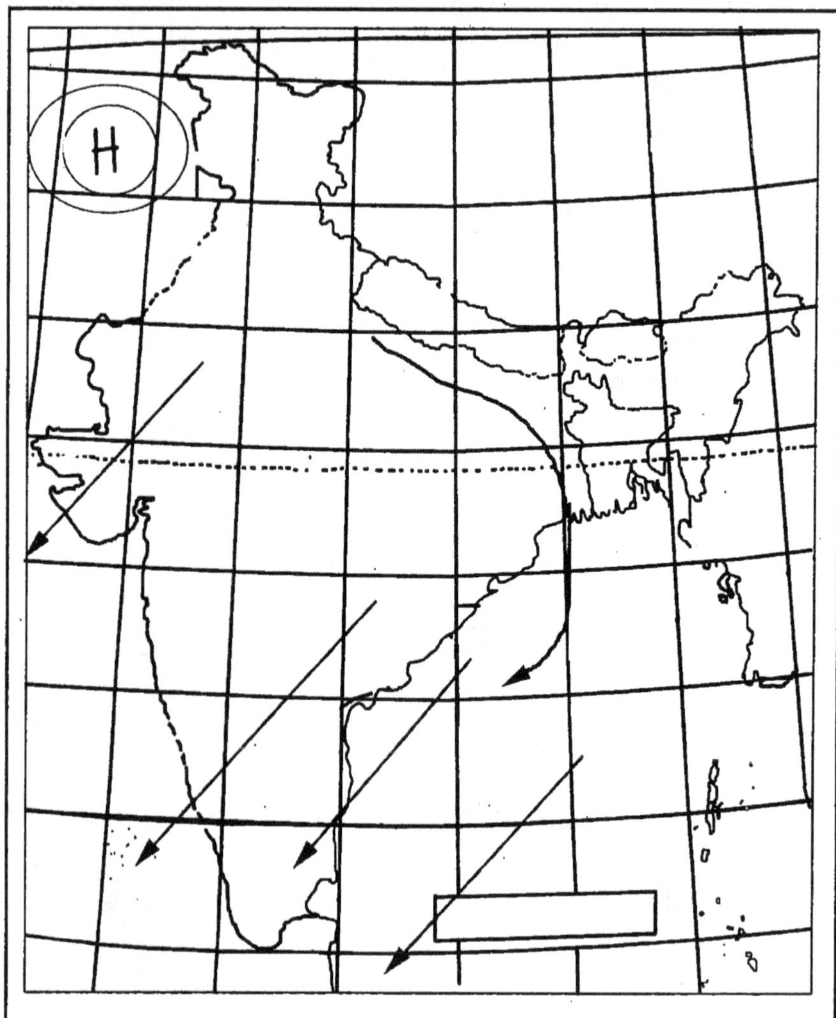

Figure : 8.7

ON SET OF NORTH EAST MONSOON

(NOVEMBER-FEBRUARY)

The driving mechanisms of monsoon

1. Differential heating of land and ocean masses causes a pressure gradient and wind is driven accordingly.
2. Twist to wind by rotation of earth.
3. Moist process determines strength, vigour, location, etc.

The path of monsoon air is distributed by diverse features

1. Earth's rotation.
2. Mountain barriers.
3. The retarding effect of friction as winds blow over land.

Withdrawal of Monsoon

1. The monsoon withdraw from northern India around mid September.
2. The monsoon withdraw from extreme south of Indian Peninsula by December.

Break and Activeness in Monsoon

1. A period of lean rainfall occurs when "Trough" shifts towards foot hills of Himalayas which is known as break in the monsoon over Indian sub-continent.
2. When the "Trough" shifts south of its normal position, monsoon becomes active over India.

Economic Importance and Influence of Monsoon Rains on Farm Operations

1. Nearly 54 per cent of population of the world depends on monsoon for their income.
2. Monsoon rains are considered as life giving rains. Rice or paddy which is a major food crop depends on only rainfall for its yield. If rainfall is not uniformly distributed, it results in huge loss of rice crop in particular and all other crops in general.
3. Heavy rain during harvesting causes lodging of crop and seed germination. If rainfall does not occur immediately after sowing, it results in germination failure.
4. As in the case of other weather elements the amount and distribution of rainfall influence the crop yield considerably.

Example : Paddy and sugarcane require high amount of water as compared to groundnut and castor.

5. Timely and evenly distributed rainfall during the crop growth is more beneficial than heavy rainfall occurring at once.
6. Rainfall of 20 m.m. is necessary to wet the soil upto a depth of 15 cm which helps in decomposition of organic matter and also influences the fertility status of the soil.
7. Many farm operations such as seed bed preparation, sowing, intercultivation, etc., depend on rainfall.

Deleterious Effects of Rainfall

1. Weeding when heavy rainfall occurs is very difficult.
2. Sowing during heavy rainfall is very difficult.
3. Rains received at the time of harvesting causes low quality seed production and some times even spoil the whole harvesting operations.
4. Much of drying of the soil surface is required for threshing to get better harvest. For instance, if rain occurs during threshing the quality of grain is affected.
5. Rain received immediately after harvesting causes germination of seeds and growth of fungal population.
6. If low rainfall occurs, it results in non-supply of sufficient water to the plant which finally results in drying up.
7. High intensity of rainfall has an adverse affect like non-availability of air to the roots, which results in death of the crop.

DROUGHT

Definition

The term drought can be defined by several ways.

1. "The condition under which crops fail to mature because of insufficient supply of water through rains".
2. "The situation in which the amount of water required for transpiration and evaporation by crop plants in a defined area exceeds the amount of available moisture in the soil".

3. "A situation of no precipitation in a rainy season for more than 15 days continuously".

The Other Causes Contributing to the Drought Condition

1. Defective tillage of soil.
2. Failure to store rain water.
3. Lack of technology with the user to retain the soil moisture.
4. High seed rate and thick plant population.

The Effects of Drought

1. Depletion of the soil moisture and reduction in ground water table.
2. Reduction of output and turnover in industry, agriculture thereby total economy of the nation.

Classification of Drought

Droughts are broadly divided into 3 categories.

1. Meteorological Drought

If annual rainfall is significantly short of certain level (75 per cent) of the climatologically expected normal rainfall over a wide area, then the situation is called by this term. In every state each region receives certain amount of normal rainfall. This is the basis for planning the cropping pattern of that region or area.

2. Hydrological Drought

This is a situation in which the hydrological resources like streams, rivers, reservoirs, lakes, wells, etc., dry up because of marked depletion of surface water. The ground water table also depletes. The industry, power generation and other income generating major sources are affected. If meteorological drought is significantly prolonged, the hydrological drought sets in.

3. Agricultural Drought

This is a situation which is a result of inadequate rainfall. As a result, the soil moisture falls short to meet the demands of the crop during its growth. Since, the soil moisture available to a crop is insufficient, it affects growth and finally results in the reduction of yield.

Some scientists consider the above classification only as a part of the total classification. The classification based on 'medium' and also 'temporal' are also in vogue.

Droughts and their Influence on Crop Plants

The influence of drought can be observed not only on phenology but also on phenophases of crop plants.

1. From seedling to ripening stages the water influence the crops particularly in case of cereals after the leaves are emerged from coleoptile. The influence of drought is more pronounced at the time of maturity.
2. During flowering stage, any little stress of moisture by virtue of drought substantially reduces the size of inflorescence thereby affecting the final yield.
3. In the same way fertilization and grain filling are also markedly influenced and the final yield is substantially reduced.
4. When soil moisture stress increases, it limits water supply to all the plant parts, which results in wilting.
5. If drought occurs at the time of grain filling, it results in the decrease of yield considerably.
6. Cell division and enlargement are very sensitive to drought stress. During moisture (drought) stress cell enlargement is affected and is the primary cause of stunted growth under field conditions.
7. Drought also affects nutrient absorption, carbohydrate and protein metabolism and translocation of ions and metabolites.
8. Protein breakdown injures the drought stressed plant due to the accumulation of toxic products such as ammonia, rather than due to a protein deficiency.
9. Abscission of leaves, fruits and seeds can be induced by plant water deficit during droughts.
10. Plant respiration is drastically reduced.

Drought Control and Management Practices

1. Modification of microclimate by use of shelter-belts and artificial barriers to reduce evapotranspiration and wind movement.
2. Maintaining optimum plant population.
3. Best possible seed-bed preparation to hold and absorb maximum moisture and better weed management.
4. Tillage practices to minimise run-off and evapotranspiration.
5. Crops that evade or endure periods of drought shall be sown.
6. Drought tolerant crops for which row spacing can be increased without affecting the final yield can be identified and practiced.
7. The dates of sowing shall be adjusted such that the reproductive stage of the crop shall not pass through the drought, in addition to other stages for critical crop growth.
8. Effective control of pests and diseases and use of recommended doses of chemicals.
9. Correcting nutrient deficiencies and use of recommended doses of fertilizers.
10. Application of anti-transpirants and use of mulches will reduce evapotranspiration.
11. Application of irrigation at appropriate stages of crop growth.
12. Weed control by keeping the land fallow has an added effect in conserving the moisture.
13. Ploughing of range lands with heavy disks or similar equipment to make a more rapid and complete infiltration.
14. Shaping of land so that the water stays where it falls or run-off from a slope to irrigate a level bench below the slope.

Analysis of Drought

Droughts can not be predicted as they are not cyclic and have no persistence. There had been attempts to analyse the drought, for many years.

1. Rainfall data alone is insufficient to quantify drought. In addition to the rainfall, air temperature was taken into consideration to classify droughts by many leading scientists in the earlier part of this century.

2. Precipitation, evapotranspiration and available soil moisture in the root zone are used as inputs to quantify the occurrence of agricultural drought.

3. Models requiring exhaustive data and calculation procedures are developed. As the crop growing season is generally lower than an year (except for horticultural plantations) it was recognised that periods of less than an year should be studied to analyse the agricultural drought.

4. Monthly weather data were used to analyse meteorological drought by many scientists. These are found to analyse the drought condition with respect to climatic aridity, agricultural production and stream flow.

5. Most of the workers interested in agricultural drought analysis now use their weekly periods or daily periods. Daily values tend to vary a great deal and are difficult to analyse on a large scale. Weekly periods are, therefore found to be more suitable in analysing drought and scheduling irrigation.

Research Needs

Research on the following topics is needed to ameliorate the conditions in drought prone areas.

1. Climatological water balance studies.
2. Dynamics of the soil moisture.
3. Delineation of arid and semi-arid regions.
4. System analysis approach for crop planning.
5. Statistical analysis of inter-relationship between various weather parameters and crop yields.
6. Water harvesting techniques and quantification of rainfall.
7. Studies on radioactive and thermal characteristics of crops.
8. Microclimatic studies like shelter-belts and wind barriers.

9. Identification of the rainfall pattern and quantification of crop yields under rainfed conditions.
10. Energy balance studies.
11. Crop weather models, etc.

Occurance of Droughts in India

1. In a period of 90 years commencing from 1765 our country experienced 12 famines of which 4 were very severe.
2. From 1862 to 1908 droughts prevailed in one part of the county or the other.
3. The worst droughts in the history of India are 1877, 1899, 1904, 1908 and 1918. Of these 1918 A.D. drought was the worst one. In this year 70 per cent of India was under drought.
4. According to the India Meteorological Department when the rainfall deficiency from the normal is 26-50 per cent, it is termed as moderate drought and when deficiency exceeds 50 per cent, it is called as severe drought.

TABLE 8.2

REOCCURRENCE OF DROUGHTS IN INDIA

S. No.	Reoccurance	Region
1.	Once in 4-5 years	Rajasthan, Gujarat, Telangana part of A.P., and Punjab.
2.	Once in 6-8 years	Haryana, Jammu and Kashmir, Coastal part of A.P., Southern part of Karnataka and Eastern part of U.P.
3.	Once in 10 years	H.P., Western U.P., Rayalaseema of A.P., and Kerala.
4.	Once in 15-20 years	Madhya Maharastra, Interior T.N., South Assam, Northern Karnataka and Orissa.
5.	Very rare	Coastal Karnataka, Bihar Plateau and Bengal.

CLOUD SEEDING

Cloud Seeding

Cloud seeding is one of the tools to mitigate the affects of drought. It is defined as, "A process in which the precipitation is encouraged by injecting artificial condensation nuclei through aircrafts or suitable mechanism to induce rain from turbulent cloud".

The rain drops are about a million times heavier than cloud drops. So, rain develops only if the cloud droplets grow by some mechanism. These mechanisms are different for cold and warm clouds.

Seeding of Cold Clouds

This can be achieved by two ways.

1. Dry Ice Seeding

- Dry ice (solid carbon-dioxide) has certain specific features. It remains as it is at – 80°C and evaporates, but, does not melt.
- Dry ice is heavy and falls rapidly from top of cloud and has no persistent effects.

Steps involved are

1. Aircraft flies across the top of a cloud and 0.5 - 1.0 c.m. size dry ice pellets are released in a steady stream.
2. While falling through the cloud a sheet of ice crystals is formed.
3. From these ice crystals rain occurs.

This method is not economical as 250 kg of dry ice is required for seeding one cloud. To carry the heavy dry ice over the top of clouds special aircrafts are required, which is an expensive process.

2. Silver Iodide Seeding

Steps involved are

1. Minute crystals of silver iodide produced in the form of smoke acts as efficient ice - forming nuclei at temperatures below -5°C.

2. When these nucleii are produced from a flame on the ground generators, these particles are fine enough to diffuse with air currents.
3. Silver iodide is the most effective nucleaturing substance because, its atomic arrangement is similar to that of ice.
4. The time taken for silver iodide smoke released from ground generator to reach the super cooled clouds was often some hours, during which it would drift a long way and decay under the sun light.
5. So, the appropriate procedure for seeding cold clouds would be to release silver iodide smoke into super cooled cloud from an aircraft.

In seeding cold clouds silver iodide technique is more useful than dry ice technique, because

(a) Very much less of silver iodide is required per cloud.

(b) There is no necessity to fly to the top of the cloud, if area to be covered is large.

Seeding of Warm Clouds

1. Water Drop Technique

1. Coalescence process is mainly responsible for growth of rain drops in warm cloud.
2. The basic assumption is that the presence of comparatively large water droplets is necessary to initiate the coalescence process.
3. So, water droplets or large hygroscopic nuclei are introduced in the cloud.
4. Water drops of 25 mm are sprayed from air craft at the rate of 30 gallons per minute and rain occurs within a few hours.

2. Common Salt Technique

1. Common salt is a suitable seeding material for seeding warm clouds.
2. It is used either in the form of 10 per cent solution or solid.

3. A mixture of salt and soap avoid practical problems.
4. The spraying is done by power sprayers and air compressors or even from ground generators.
5. The balloon burst technique is also beneficial. In this case gun powder and sodium chloride are arranged to explode near cloud base dispersing salt particles.

Results of Research on Effect of Rainfall on Important Crops

I. Rice

1. Under rainfed rice cultivation, where temperatures are within the critical ranges, rainfall is the most critical factor limiting rice cultivation.
2. Rainfed rice cultivation is limited to areas where annual rainfall is more than 1000 milli metres.
3. The variation in amount and distribution of rainfall is the most important factor limiting yields of rainfed rice which constitute about 80 per cent of rice grown in south and south-east Asia.
4. The amount and distribution pattern of rainfall varies widely from location to location and year to year.

   Example : The Deccan Plateau of Indian sub-continent receives a rainfall of 500 millimetres, which is much less than the water requirement of rice crop (1240 millimeters). Hence, rice is grown when irrigation is given. But, in Philippines where 2468 m.m. of rainfall is received per year, from total monsoon (SW and NE monsoons), rainfall is fairly and evenly distributed throughout the year. There is no dry season. Hence, rice is grown throughout the year, but harvesting and drying are the problems. Whereas, Cuttack in India, receives 1545 m.m. per year from south-west monsoon only. Rainfall is distributed from June to November evenly and this crop is grown favourably.

5. The effects of water shortage and excess water are similar to any other crop depending upon the stage of the crop like leaf rolling, leaf scorching, impaired tillering, stunting, delayed flowering, spikelet sterility and incomplete grain filling. Rice crop is most sensitive to water deficit from reduction division to heading stages. Presence of drought for 3 days before heading reduced the yields significantly causing 59 to 62 per cent spikelet sterility.

II. Groundnut

1. Rainfall is the most important weather parameter which not only influences vegetative but also reproductive stages.
2. Erratic behavior of rainfall i.e., low amount that too highly variable coupled with unfavorable conditions are the main causes of low yields in light soils with low water holding capacity.
3. The ideal pattern of rainfall distribution for groundnut is pre-sowing 80 - 120 millimetres, at sowing 100 - 120 millimetres flowering to peg penetration 200 millimetres and pod development and pod maturation 200 millimetres.
4. A temporary stress of 15-20 days after the stand establishment is better to check the excessive vegetative growth and to promote flowering.
5. Prolonged stress reduces the uptake of NPK, delays flowering, reduction in number of flowers, flowering period, flower opening, pollen viability, etc.

III. Sugarcane

1. This crop is grown from 750 m.m. to 2500 m.m. rainfall areas. In low rainfall areas the crop is grown as dry and depends for moisture on irrigation, and under high rainfall conditions it is cultivated as rainfed crop.
2. During the formative phase high amounts of rainfall are desirable. As the crop advances the water requirements are reduced. Under more rainfall the pests and diseases effect the crop.

IV. Cotton

1. Cotton is grown in arid climates and it prefers moderate rain to excessive moisture. Heavy rains injure the young seedlings and often harm the fully grown plant. They also cause the lodging of plants.
2. For successful cultivation of cotton even distribution of rainfall is of far greater importance than the total annual amount.
3. In regions with inadequate rainfall, yield is substantially increased by irrigation.

Chapter - 9

Weather Disaster Management, Synoptic Reports, Weather Forecasting and Remote Sensing

"It is said that one result is obtained by worshiping the supreme cause of all causes and that another result is obtained by worshiping what is not supreme. All this is heard from the undisturbed authorities, who clearly explained It".

WEATHER DISASTER MANAGEMENT

Crops depend upon certain optimum weather conditions for their potential production, although other variables such as fertilizers, insecticides, etc., interact to certain extent in an agricultural system. Daily, seasonal and long term variations in any or all the climatic elements alter the efficiency of plant growth thereby the crop production. The deviation of climatic factors considerably from their normal values is referred as the "Adverse weather" or the "Adverse climate" depending on duration of such impact.

The following are the adverse weather conditions and the possible management strategies.

I. Rainfall

The rainfall is the major source of water which is essential for plant growth and development. However, the rainfall is considered adverse, if it is

(a) Excess rainfall (b) Scanty rainfall (c) Untimely

The total amount of rainfall in a season is not the criteria. But, its well distribution over a large area is desirable. Heavy rains with short frequencies will result in floods. If 125 mm of rain is received in two and half hours it is called as heavy rain.

A. Excess Rainfall

1. Eventhough, water in all its forms play a fundamental role in the growth and production of all crops, excessive amounts of water in the soil alter various chemical and biophysical processes.
2. Free movement of oxygen is blocked and compounds toxic to the roots are formed due to drainage problem.
3. Soils with high rate of percolation are unsuitable for cultivation as plant nutrients can be removed rapidly.
4. Heavy rains directly damage plants on impact or interfere with flowering and pollination.
5. Top soil layers are packed or hardened which delays or prevents emergence of tender seedlings.
6. Snow and freezing rain are threats to winter plants. The sheer weight of ice and snow may be sufficient to break limbs on trees and shurbs.
7. A thick ice cover on the ground tend to produce suffocation of crop plants such as winter wheat.
8. Under excess rainfall conditions floods occure in areas drained by large river systems.
9. Floods submerge crops, silt up fields, tank bunds and river embankments are washed off.

Management of Excess Rainfall (Floods)

1. Construction of multipurpose projects such as irrigation and electric systems.
2. Planned afforestation.
3. Keeping the field drains open.
4. Growing flood obstructing crops.

B. Scanty rainfall

This is a synonym with "Inadequate rainfall" or "Drought". The influence of drought can be observed not only on phenology but also on phenophases of crop plants (chapter 8).

1. Water limitation from seedling emergence to maturity in all the cereals is very damaging.
2. Water stress/drought during flowering reduces the size of inflorescence, effect fertilization, grain filling and reduce final yield.
3. Plants show wilting symptoms.
4. Cell division and enlargement are very sensitive to drought stress, which results in stunted growth.
5. Drought effects nutrient absorption, carbohydrate and protein metabolism and translocation of ions and metabolites.
6. Abscission of leaves, fruits and seeds can be induced by plant water deficit during droughts.
7. Plant respiration is drastically reduced.

Management of Drought / Scanty Rain

1. Application of sufficient irrigation water negates the condition of insufficient or scanty rain.
2. Discover amount of water needed at various stages and adjust the sowing dates.
3. Conserve water by suitable management of fallow and cropped fields viz., breaking up the surface to reduce runoff, removal of weeds, digging pits of small size which collects runoff water, etc.

C. Untimely Rains

This refers to rainfall received too early or too late in the season with the result that normal agricultural operations are upset (chapter 8).

1. Too early rains do not permit proper preparation of seedbed due to slushyness.
2. Too late rains delay sowings and pest attack causes collosal losses.
3. Wet spells during flowering and harvesting results in poor fertilization and subsequent loss in yield.

Management of Untimely Rains

1. Farmers shall be advised to follow the weather forecasting by IMD for proper management of their crops through crop-weather advisories.
2. Contingency crop plans shall be made available to the needy farmers.

II. Temperature

Temperature is essential for all plant physiological processes, gaseous exchange between plant and environment, stability of plant enzymatic reactions, etc (chapter 3). However, both cold and heat waves and abnormal soil temperatures are adverse to crop growth and development.

A. Cold Waves

During winter (December - February), temperature decreases generally over the Indian sub-continent. It is lower in northern India and higher in southern India. This fall in temperature may cause damage to the crops. If the temperature drops to freezing or below, a frost may occur which causes severe damage to the crops/crop plants. Threat of frost is danger to crops. Forst is a form of condensation that forms on cold objects when the dew point is below freezing (chapter 8). Frosts are of two types :

(a) ***Advection or airmass frost :*** This results when the temperature at the surface in an airmass is below freezing.

(b) ***Radiation frost :*** This occurs on clear nights with a temperature inversion.

3. There is a special case of frost caused by loss of heat by evaporation. This occurs when cold rain showers wet the leaves and are then followed by the dry wind.

(a) *Advection Frost*

The usual effects of advection frost are :

1. The injury and death caused by frost is due to the formation of ice crystals in and outside the plant cells.
2. During dormancy, plants can withstand lower temperatures upto -20°C.
3. Once growth has commenced temperatures of few degrees below freezing point may be fatal.
4. The cell sap gets frozen below 0°C, as also between cells.
5. Extra cellular ice formation occurs followed by withdrawl of water from the cell.
6. The protoplasm may become dehydrated and brittle, resulting in mechanical damage or the cell may contract and damage the protoplasm.

Management of Advection Frost

For production of most field crops, the only satisfactory solution to the problem of advection freezing is to avoid it as far as possible by planting after the damage is past and by selecting varieties which will mature before the beginning of the hazard.

(b) Radiation frost

The damage due to radiation frost differs from the above freeze damage in degree and its spotly occurrence.

1. This radiation frost damage is critical during critical stages of growth.
2. Young seedlings may be killed.
3. Flowering stage is most prone.
4. Crops like potato, tomato and melons are vulnerable right upto maturity.

5. For most field crops and orchard crops flowering stage is most critical for frost damage.

6. Forsty nights followed by warm sunny days produce a sunclad on orchard fruits, considerably reducing their production.

Management of Radiation Frost

The management of radiation frost can be grouped into passive and active methods.

Passive methods

- Clean cultivation.
- Maintenance of the soil moisture.
- Wrapping plants with insulating material and enclosing the basal part of the plant.
- Proper site selection.
- Choice of growing season.
- Breeding of cold resistant varieties.

The above methods can be followed even for advection frost also. These passive methods do not involve any modification of environment.

Active methods

The active methods of frost protection are many, like use of

- Heaters.
- Wind machines.
- Sprinkling water.
- Following weather forecasst for better management of crops.

B. Heat Waves

These are very harmful during the summer. These are experienced over the Deccan and Central parts of India from March to May. The harmful effects include :

1. Shedding of fruits, leaves, drying of water resources, etc.

2. Loss of water by evaporation from irrigation channels.

3. Transpiration increases from plants beyond recouping levels.
4. Plants tend to wilt and die owing to rapid dessication.
5. Hot winds cause shrivelling effect at milk stage of all agricultural crops.

Management of Heat Waves

Adoption of specific agronomic practices like, shelterbelts, wind breaks, choice of varieties, etc.

III. Wind

The wind has its most important effects on crop production indirectly through the transport of moisture and heat. Vegetative growth at "Zero wind", as experienced in glass houses or under low glass cover is luxurient. But, there is typically a reduction in vegetative growth as the wind increases to small values, viz., 1 or 2 metres per second.

A. Beneficial Effects of Winds

1. Moderate turbulence promotes the consumption of carbon-dioxide by photosynthesis.
2. Prevent frost by disrupting a temperature inversion.
3. The wind dispersal of pollen and seeds is natural and necessary for certain agricultural crops and natural vegetation also.

B. Harmful Effects of Winds

1. At sustained high speeds (12-15 metres per second) at plant height, plants assume a low, dwarf like form, whilest the intermittent high wind speeds experienced in gales, hurricanes, etc., results in gross physical damage to bushes and trees.
2. At higher wind speeds, the shape of the orchard tree alters giving rise to the characteristic wind shaping of trees in exposed positions.
3. Leaves become smaller and thicker.
4. Breakage occurs, bushes and trees are subjected to natural (seasonal) pruning.
5. Direct mechanical effects are the breaking of plant structures, lodging of cereal crops or shattering of seed from panicles.

Management of High Winds

1. The effects of wind on evaporation can be avoided by using proper method of irrigation.
2. The damaging effect of wind can be reduced over a limited area by the use of shelter belts (rows of trees planted for wind protection) and wind breaks (any structure that reduce the wind speed).

IV. Thunderstorms, Dust Storms and Hail Storms

These storms are known as local severe storms. As many as 44,000 thunder storms occur daily on earth.

1. These are formed in a situation where a great deal of the energy for their genesis and development comes from the release of the latent heat of condensation in rising humid air.
2. These local storms cause severe damage to the standing crops through mechanical injury to the plants.
3. In dust storms, the dust rised by the wind covers small plants, which may cause stomata closure and suffocation.
4. Hails cause direct damage to crops by lodging, shattering of seeds, etc., depending on their intensity.

Management of Storms

1. Prevention of hails by hail suppression techniques.
2. Following forecasts of weather and protecting crops.
3. Spraying of salt on harvested paddy, to prevent the germination and sprouting of the harvested produce.

V. Excessive or Defective Insolation

Excessive solar radiation results in rise of the soil and air temperatures. Defective insolation with consistantly cloudy weather on one hand and consistantly bright and high intensity sunshine on the other hand causes enormous damage to crop plants.

1. Cloudy weather retard growth, affects pollination, causes disease and pest incidence.
2. High solar radiation intensity causes pollen burst or flower drop.

Management

Since, these are very rare, the location specific solutions are feasible.

1. Proper site selection.
2. Allowing air drainage.
3. Adequate water supply.
4. Pruning of orchard trees.
5. Spray of chemicals and plant harmones.
6. Covering plants with "Hot caps" (covering plants with some standard and recommended material).

VI. Tornado

This is a violent, destructive storm of small horizontal dimensions. A cumulonimbus cloud forms into a funnel shape with an vortex extending from the base of the storm to the surface. The whirl-wind encircles a small dimension of about 500 metres. These are capable of causing severe structural and other damages. The violent winds associated with this abnormality are strong upward air currents. The tornados occurring on water are known as the "Water spouts".

Management

1. Warning in advance
2. Precautions to protect the agricultural produce like covering the harvested produce, transportation to safety places, etc.,
3. Quick removal of debris immediately after damage.

SYNOPTIC REPORTS

For better crop management under adverse weather conditions synoptic climatology play an important role.

The term synoptic climatology is applied to investigations of regional weather and circulation types. It is also used to refer to any climatological analysis which makes some reference to synoptic weather phenomena.

This field is concerned with obtaining an insight into local or regional climates by examining the relationship of weather elements individually or collectively to atmospheric circulation processes.

Synoptic climatology is defined as, "The description and analysis of the totality of weather at a single place or over a small area, in terms of the properties and motion of the atmosphere over and around the place or area".

There are essentially two stages to a synoptic climatological study.

1. The determination of categories of the atmospheric circulation type.
2. The assessment of weather elements in relation to these categories.

Besides agricultural meteorological observatories, synoptic weather stations also record weather data such as rainfall, temperature, radiation, low level wind, evaporation, etc. The surface observatories collect information on various weather elements and based on these recordings daily forecasts, warnings and weather reports are prepared by 5 regional forecasting centres at Chennai, Nagpur, Mumbai, Delhi and Kolkata. The weather bulletins are being broadcast in regional languages through the All India Radio and Television.

Synoptic Report

Observed weather conditions are marked in brief coded form as a synopsis of the conditions. Such a brief report on weather conditions is known as the "Synoptic report".

Synoptic Chart/Weather Map

The regular observatories record weather elements at scheduled time and send these readings through a telegram to the main observatory at Pune. They reach Pune within an hour of observation and they are charted on outline map of India, using the international code of signals and abbreviations. These are called the "Synoptic charts or the weather maps".

In the synoptic charts different weather phenomena and atmospheric characters are marked with different symbols (Table 9.1).

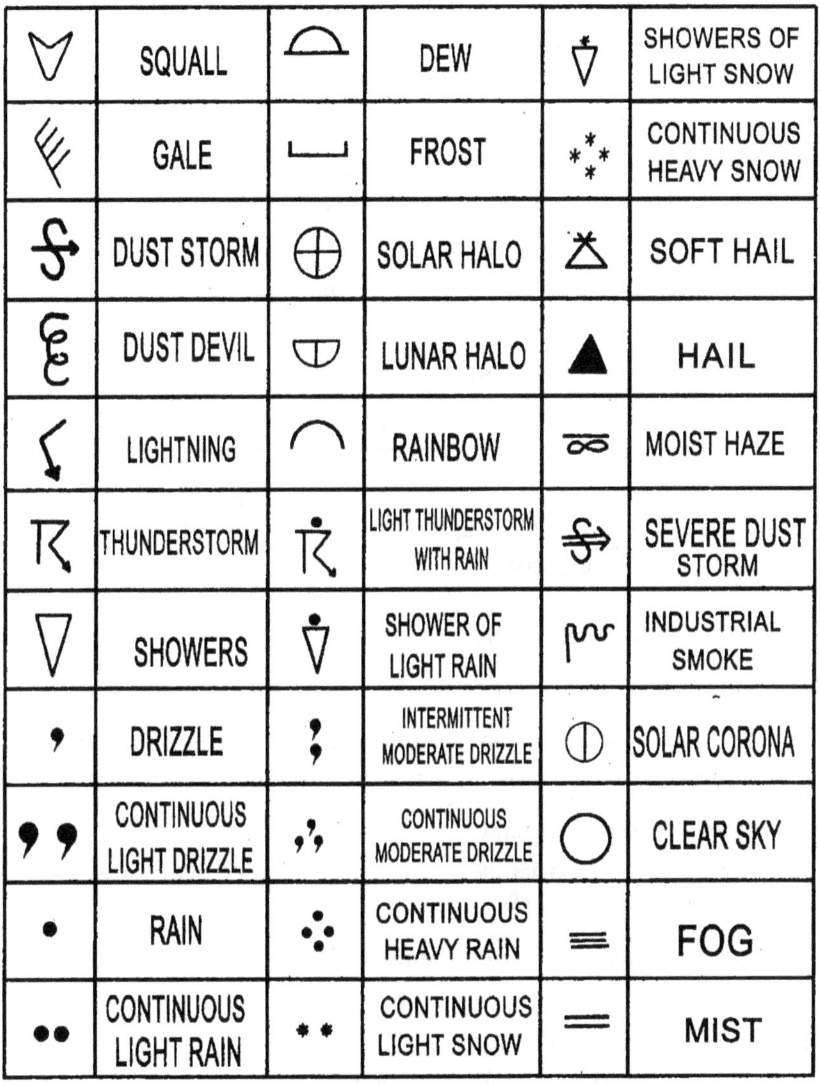

SQUALL	DEW	SHOWERS OF LIGHT SNOW
GALE	FROST	CONTINUOUS HEAVY SNOW
DUST STORM	SOLAR HALO	SOFT HAIL
DUST DEVIL	LUNAR HALO	HAIL
LIGHTNING	RAINBOW	MOIST HAZE
THUNDERSTORM	LIGHT THUNDERSTORM WITH RAIN	SEVERE DUST STORM
SHOWERS	SHOWER OF LIGHT RAIN	INDUSTRIAL SMOKE
DRIZZLE	INTERMITTENT MODERATE DRIZZLE	SOLAR CORONA
CONTINUOUS LIGHT DRIZZLE	CONTINUOUS MODERATE DRIZZLE	CLEAR SKY
RAIN	CONTINUOUS HEAVY RAIN	FOG
CONTINUOUS LIGHT RAIN	CONTINUOUS LIGHT SNOW	MIST

Figure:9.1

SYMBOLS FOR RECORDING WEATHER PHENOMENA

TABLE 9.1

CHARACTER OF SYMBOLS USED IN SYNOPTIC CHARTS

S. No.	Symbols	Weather element/character/ phenomenon
1.	Narrow black lines	Isobars
2.	Numbers at ends of isobars	Pressure values in millibars.
3.	Shading	Precipitation
4.	Arrows	Wind direction
5.	Feathers in the arrows	Wind velocity
6.	Small circles with shading	Amount of clouds

In addition to the above, different symbols (Figure 9.1) are used for recording weather phenomena, in relevant columns of the pocket register and the monthly meteorological register by the observer.

The Duties of the Observer

The routine duties of the observer include :

1. To make regular and careful observations and to note the general character of the weather and record in the pocket register.
2. To prepare and dispatch the weather telegram as per the instructions to the different forecasting centres, immediately after the observations are taken.
3. To send heavy rainfall telegrams to the various offices on warning list.
4. To prepare and post monthly meteorological and pocket registers for each month to the controlling meteorological office.
5. To keep the instruments clean and maintain them properly.

After the observer sends the data as per the standard procedure it should be decoded and the weather observations for each station must be plotted at the appropriate location in a systematic manner following the international station model. Only weather maps in first class forecasting centres approach the completeness of this model. Printed maps and maps used for plotting, usually have an appropriately numbered circle corresponding to each reporting land station and observations are plotted about this location in the appropriate position regardless of the number of observations shown. The weather pattern affecting a locality is an integral part of the much larger hemispheric weather pattern and it is necessary to plot a map over a large area. Even if, observations are not to be plotted, it is necessary to know the plotting scheme in order to read and interpret weather charts already plotted.

WEATHER FORECASTING

The weather elements which influence the agricultural operations and crop production can be forecast upto different spans of time. Weather forecast is defined as, "The prediction of weather for the next few days to follow".

Utility of Weather Forecasts

In India the total annual pre-harvest losses of various crops range from 10 to 100 per cent. Similarly, the post-harvest losses average upto 10 per cent. When an accurate weather forecast is given for the needs of agriculture it contributes immensely to the monetary benefits of the farmers.

1. Short term adjustments in daily and weekly agricultural operations.

   Example : If heavy rain occurs immediately after sowing of seeds the seeds are washed away. If a hail storm occurs during harvesting it causes shedding of grains and fruits. If warned in time the farmer would hurry up some of the operations or postpone them suitably adjusting the cropping operations to weather conditions.

2. Minimising input losses resulting from adverse weather (seeds, chemicals, fertilizers, diesel or electric power used for irrigation, etc.).

Example : The yield of the crop is determined by weather conditions to a greater extent. If weather is predicted in advance the amount spent on irrigation, electricity, labour, etc., can be reduced substantially. Nearly, 50 per cent of farmers will definitely be benefited if warnings are given well in advance to them. The critical periods for normal growth of the crop can be adjusted for healthy growth and development of crop.

3. Markedly improve the yields of crops both qualitatively and quantitatively.

   Example : If farmer knows when the monsoon rains are likely to commence and how the rainfall could be from time to time in the season he would be able to plan his agricultural operations like preparation of seed bed, manuring, intercultivation including drying and threshing of the produce.

Prime Requirements for Weather Forecasting

1. A good data set.
2. A good method which can be used to forecast.

Weather Data Used in Forecasting

The following two sets of weather elements are measured routinely :

1. Pressure, temperature, wind (speed and direction) and humidity.
2. Rainfall, cloud (type and amount), visibility, pressure change, present and past weather, maximum and minimum temperatures, etc.

Types of Observations

The main observations used in different weather forecasting types are as follows :

1. Surface observations
2. Upper air observations
3. Aircraft observations
4. Radar observations

Types of Weather Forecast

There are three types of weather forecast (Table 9.2) which are commonly followed worldwide.

TABLE 9.2

TYPES OF WEATHER FORECAST AND THEIR VALIDITY

S. No.	Type of forecast	Validity period	Main users	Predictions
1.	Short range (a) Now casting (b) Very short range	Upto 72 hours 0-2 hours 0-12 hours	Farmers, marine agencies, general public, etc.	Rainfall distribution, heavy rainfall, heat and cold wave conditions, thunder storms, etc.
2	Medium range	Beyond 3 days and upto 10 days	Farmers	Occurrence of rainfall, temperature intensity, etc.
3.	Long range	Beyond 10 days; a few weeks to a month, and season	Planners	This forecast is provided for Indian monsoon rainfall. The out looks are usually expressed in the form of expected deviation from normal conditions.

Methods Used in Weather Prediction

Three methods are used for accurate weather prediction.

1. Synoptic Method

This is a subjective technique. In this method weather charts are analysed and the analogous situations happened in the past are matched with the present situation. This method is useful for the present situation (for short range forecast). The success of the forecast depends on the skill and experience of the forecaster.

2. Statistical Method

In this method correlations and regressions are calculated using weather elements. This method is useful for long range weather forecast.

3. Numerical Method

This is basically an objective technique. Several equations are solved numerically using high speed and large memory computers. This method is useful for short and medium range forecasts.

Agromet Advisories

1. The IMD has established Agromet Advisory Service Units (AASUs) at the meteorological offices of the state head quarters. These AASUs issue bi-weekly agromet advisories to the states. First, the condition of the crops in the state and then advisory on the farming operations, based on the past weather / likely future weather realised is provided. The rainfall forecasts valid for the next two days and the out look valid for two subsequent days is also given. Assistance for the agricultural related aspects is taken from the state agricultural universities and agricultural departments.
2. The agromet advisories are sent to the "Farm radio" division of the All India Radio stations through land line telegrams and are broadcast in the farm radio programmes of respective states. A separate pictorial presentation of spatial rainfall distribution over the state is sent to the Doordarshan for telecasting in the respective states.
3. The advisories are also sent through fax to the Agromet Directorate of IMD, Pune on the same day, where all the advisories sent by the various AASUs are assembled and then a consolidated report is prepared. This report is faxed from IMD Pune to IMD Delhi on subsequent day and is used for ministerial/secretarial briefing.

REMOTE SENSING

The word "Remote sensing" was coined by Fischer in 1960 AD. Remote sensing is defined as, "The collection and interpretation of information about a target without being in physical contact with it". According to Lilesand and Kiefer, remote sensing is, "The science and art of obtaining information about an object, area or phenomenon through the analysis of data acquired by a device that is not in contact with the object, area or phenomenon under investigation".

Basic Concepts of Remote Sensing

There are two basic interactions between electromagnetic energy and earth surface features. These interactions are considered as basic concepts of remote sensing.

1. The propositions of energy reflected, absorbed and transmitted will vary for different earth features, depending on their material type and condition. These differences permit to distinguish different features on an image.
2. Even within a given feature type, the propotion of reflected, absorbed and transmitted energy will vary at different wavelengths.

Approaches for Information

Mainly, there are three approaches to obtain information.

1. Visual interpretation.
2. Digital image processing.
3. Satellites and sensors.

Platforms used in Remote Sensing

Taking aerial photographs by camera, tracking any object or phenomenon by radar, seeing the object by our eyes or producing satellite images are the examples of remote sensing. Basically, there are three platforms, used in remote sensing.

1. Ground based remote sensing

If the platform having sensors is at the ground it is known as ground based remote sensing.

2. Air borne remote sensing

If the platform is upto 100 kilometres height with smaller coverage capability like the aircrafts, balloons or rockets it is called as air borne remote sensing.

A remote sensor can detect variation in reflectance between objects depending upon four interrelated factors.

(a) The radiometric resolution of the sensor.

(b) The amount of atmospheric scatter.

(c) The surface roughnes of the objects.

(d) The spectral variability of reflectance within the scene.

3. ***Space borne remote sensing***

If the platform is in the space having capability of global coverage it is called as space borne remote sensing.

Multispectral Scanning

The generalised processes and elements involved in electromagenetic remote sensing of earth resources are :

(a) Data acquisition.

(b) Data analysis.

Sensors are used to record variations in which earth surface features reflect and emit electromagnetic energy. Multispectral scanners use sensors with very narrow fields of view to scan an area of interest systematically and an image is built up as the scan progresses. Each sensor forms an image that represents the reflectance of the scan in its particular wavelength of the scene in its particular waveband. The Indian Remote Sensing Satellite System provides images with 36.25 m and 72.5 m ground resolution in the bands 0.45 – 0.52 microns; 0.52 – 0.59 microns; 0.62 – 0.68 microns and 0.77 – 0.86 microns.

Spectral Reflectance for Vegetation, Soil and Water

In the field of agriculture the scientists are interested in spectral signatures (characteristics) of vegetation, soil and water.

These are useful to describe the nature of energy – matter interaction when the energy is in; (a) Visible (b) Near infrared and (c) Middle infrared wavelength bands.

Visible and adjacent infrared spectrum provides useful information for the study of plant communities. Within this range spectral characteristics of soil, water and vegetation differ significantly for different spectral bands.

1. Vegetation

- In visible region chlorophyll absorption bands are approximately central at 0.45 to 0.65 microns.
- Internal structure of leaf is different for different vegetations and this difference gets noticed in near infrared to mid-infrared spectral reflectance.
- Crop discrimination gets improved with the measurements in 1.55 to 1.75 and 2.05 to 2.35 micron bands which are sensitive to crop moisture content or leaf air space volume

2. Soil

Soil shows gradual increase in reflectance as the wavelength of optical spectrum increases.

3. Water

- Water bodies show fairly good response to shortwave radiation.
- It also shows absorption in infrared bands.

Applications of Remote Sensing in Agriculture

Remote sensing is an effective tool in assessing the damages to crops and their management.

1. Monitoring In-season Agricultural Operations

All the farm operations like sowing, inter-cultivation, harvesting, etc., is being monitored effectively by the remote sensing.

2. Crop Identification

By using LISS II or III sensors crop identification on regional scale is possible.

3. Crop Acerage Estimation

By using stratified sampling methodology crop acerage estimation is done to the high level precision.

4. Crop Yield Estimation

The crop yields are estimated by analysing satellite based vegetation indices which are transformations of reflectance in the near infrared portions of electromagnetic spectrum.

5. Monitoring of Crop Phenology and Stresses

The crop condition is affected by several factors like deficiency of nutrients, acidic and salinity problems of soil, nutrient deficiencies, adverse weather conditions, etc. All these can be detected by remote sensing.

6. Damage Assessment and Command Area Management

The damages due to floods, cyclones, water logged areas in command area, etc., can be detected and managed effectively by using techniques like the Multi-Temporal Remote sensing.

7. Water Availability and Soil Moisture Estimation

The surface and sub-surface water availability for irrigation and the amount of moisture stored in the upper few centimetres of the soil can be found with a greater accuracy.

8. Land Degradation and Watershed Management

The remote sensing technology is highly useful in identifying and delineating degraded lands. Also, it facilitates in delineation of the watershed areas.

9. Drought Detection and Management

Assessing the drought realistically and ways to manage the adverse effects is possible through remote sensing.

10. Desertification

Remote sensing provides information to identify the important indicators of desertification. Based on this, action can be taken by the planners at different levels.

Indian Remote Sensing Satellites

The Indian Government has launched several satellites (Table 9.3) for the benefit of general public and farmers.

TABLE 9.3

MAJOR INDIAN SATELLITES AND THEIR DATES OF LAUNCH

S.No.	Name of the satellite	Altitude	Launch Date
1.	IRS-1A	904	17-03-1988
2.	IRS-1B	904	28-08-1991
3.	IRS-P2	817	15-10-1994
4.	IRS-1C	817	28-12-1995
5.	IRS-P3	737 km (P)	
		821 km (A)	21-03-1996
6.	IRS- 1D	-- do --	29-9-1997

P = Perigee; A = Apogee.

Chapter - 10

Crop Growth Modelling, Climate Change and Climatic Classification

"One should know perfectly the Personality of God-head and His transcendental name, form, qualities and pastimes, as well as the temporary material creation with its temporary demigods, men and animals. When one knows these, he surpasses death and the ephemeral cosmic manifestation with it, and in the eternal kingdom of GOD he enjoys his eternal life of bliss and knowledge".

CROP GROWTH MODELLING

Crop is defined as, "Aggregation of individual plant species grown in a unit area for economic purpose".

Growth is defined as, "Irreversible increase in size and volume and is the consequence of differentiation and distribution occurring in the plant".

Simulation is defined as, "Reproducing the essence of a system without reproducing the system itself". In simulation the essential characteristics of the system are reproduced in a model which is then studied in an abbreviated time scale.

A model is a schematic representation of the conception of a system or an act of mimicry or a set of equations which represent the behaviour of a system. Also, a model is a representation of an object, system or idea in some form other than that of the entity itself. Its purpose is usually to aid in explaining, understanding or improving performance of a system. A model of an object may be an exact replica of the object or it may be an abstraction of the object's salient properties. A model is, by definition "A simplified version of a part of reality, not a one to one copy". This simplification makes models useful because it offers a comprehensive description of a problem or a situation. However, the simplification is, at the same time, the greatest drawback of the process. It is a difficult task to produce a comprehensible, operational representation of a part of reality, which grasps the essential elements and mechanisms of that real world system and even more demanding, when the complex systems encountered in environmental management.

The Earth's land resources are finite, whereas the number of people that the land must support continues to grow rapidly. This creates a major problem for agriculture. The production must be increased to meet rapidly growing demands while natural resources must be protected. New agricultural research is needed to supply information to farmers, policy makers and other decision makers on how to accomplish sustainable agriculture over the wide variations in climate around the world. In this direction explanation and prediction of growth of managed and natural ecosystems in response to climate and soil - related factors are increasingly important as objectives of science. Quantitative prediction of complex systems, however, depends on integrating information through levels of organization, and the principal approach for that is through the construction of simulation models. Simulation of system's use and balance of carbon, beginning with the input of "Carbon" from canopy assimilation forms the essential core of most simulations that deal with the growth of vegetation.

Systems are webs or cycles of interacting components. Change in one component of a system produces changes in other components because of the interactions. For example, a change in weather to warm and humid may lead to the more rapid development of a plant disease, a loss in yield of a crop, and consequent financial adversity for individual farmers and so for the people of a region. Most natural systems are complex. Many do not have boundaries. The bio-system is comprised of a complex interaction among the soil, the atmosphere, and the plants that live in it. A chance alteration of one element may yield both desirable and undesirable consequences. Minimizing the undesirable, while reaching the desired end result is the principal aim of the agrometeorologist. In any engineering work related to agricultural meteorology the use of mathematical modelling is essential. Of the different modelling techniques, mathematical modelling enables one to predict the behaviour of design while keeping the expense at a minimum. Agricultural systems are basically modified ecosystems. The problems of managing these systems are very difficult. These systems are influenced by the weather both in length and breadth. So, these have to be managed through systems models which are possible only through classical engineering expertise.

Types of Models

Depending upon the purpose for which it is designed the models are classified into different groups or types.

(a) Statistical Models

These models express the relationship between the yield or yield components and the weather parameters. In these types relationships are measured in a system using statistical techniques.

Example : Step down regressions, correlation, etc.

(b) Mechanistic Models

These models explain not only the relationship between the weather parameters and the yield, but also the mechanism of these models explains the relationship of influencing dependent variables.

(c) Deterministic Models

These models estimate the exact value of the yield or dependent variable. These models also have defined co-efficients.

(d) Stochastic Models

A probability element is attached to each output. For each set of inputs different outputs are given alongwith probabilities. These models define the yield or state of dependent variable at a given rate.

(e) Dynamic Models

Time is included as a variable. Both dependent and independent variables are having values which remain constant over a given period of time.

(f) Static Models

Time is not included as a variable Dependent and independent variables having values remain constant over a given period of time.

(g) Simulation Models

Computer models, in general, are a mathematical representation of a real world system. One of the main goals of crop simulation models is to estimate agricultural production as a function of weather and soil conditions as well as crop management. These models use one or more sets of differential equations, and calculate both rate and state variables over time, normally from planting until harvest maturity or final harvest.

(h) Descriptive Models

A descriptive model defines the behaviour of a system in a simple manner. The model reflects little or none of the mechanisms that are the causes of phenomena but, consists of one or more mathematical equations. An example of such an equation is the one derived from successively measured weights of a crop. The equation is helpful to determine quickly the weight of the crop where no observation was made.

(i) Explanatory Models

These consist of quantitative description of the mechanisms and processes that cause the behaviour of the system. To create this model, a system is analysed and its processes and mechanisms are quantified separately. The model is built by integrating these descriptions for the entire system. It contains descriptions of distinct processes such as leaf area expansion, tiller production, etc. Crop growth is a consequence of these processes.

IBSNAT and DSSAT

In many countries of the world, agriculture is the primary economic activity. Great numbers of the people depend on agriculture for their livelihood or to meet their daily needs, like food. There is a continuous pressure to improve agricultural production due to staggering increase in human population. Agriculture is very much influenced by the prevailing weather and climate. The population increase is 2.1 per cent in India and this is almost same throughout the world. This demands a systematic appraisal of climatic and soil resources to recast an effective land use plan. More than ever farmers across the globe want access to options such as the management options or new commercial crops. Often, the goal is to obtain higher yields from the crops that they have been growing for a long time. Also, while sustaining the yield levels they want to :

1. Substantially improve the income.
2. Reduce the soil degradation.
3. Reduce dependence on off-farm inputs.
4. Exploit local market opportunities.
5. Farmers also need a facilitating environment in which;
   (a) Affordable credit is available.
   (b) Policies are conducive to judicious management of natural resources.
   (c) Costs and prices of production are stable.
6. Another key ingredient of a facilitating environment is INFORMATION, such as :
   (a) An understanding of which options are available.
   (b) How these operate at farm level.
   (c) The impact on issues of their priority.

To meet the above requirements of resource poor farmers in the tropics and sub-tropics IBSNAT (International Benchmark Sites Network for Agro-technology Transfer) began in 1982. This was under a contract from the U.S. Agency for International Development to the University of Hawaii at Manoa, USA. IBSNAT was an attempt to demonstrate the effectiveness of understanding options through systems analysis and simulation for ultimate benefit of farm households across the globe. The purposes defined for the IBSNAT project by its technical advisory committee were to :

1. Understand ecosystem processes and mechanisms.
2. Synthesize from an understanding of processes and mechanisms, a capacity to predict outcomes.
3. Enable IBSNAT clientele to apply the predictive capability to control outcomes.

The models developed by IBSNAT were simply the means by which the knowledge scientists have and could be placed in the hands of users. In this regard, IBSNAT was a project on systems analysis and simulation as a way to provide users with options for change. In this project many research institutions, universities, and researchers across the globe spent enormous amount of time and resources and focussed on :

1. Production of a decision support system capable of simulating the risks and consequences of alternative choices, through multi-institute and multi-disciplinary approaches.
2. Definition of minimum amount of data required for running simulations and assessing outcomes.
3. Testing and application of the product on global agricultural problems requiring site-specific yield simulations.

The major product of IBSNAT was a Decision Support System for Agro-Technology Tranfer (DSSAT). This was developed by the network members led by Prof.J.W.Jones, distinguished Professor, Department of Agricultural and Biological Engineering, University of Florida, Gainesville, USA. The DSSAT is being used as a research and teaching tool. As a research tool its role to derive recommendations concerning crop management and to investigate environmental and sustainability issues is unparalleled. The DSSAT products enable users to match the biological requirements of crops to the physical characteristics of land to provide them with management options for improved land use planning. The DSSAT is being used as a business tool to enhance profitability and to improve input marketing.

The traditional experimentation is time consuming and costly. So, systems analysis and simulation have an important role to play in fostering this understanding of options. The information science is rapidly changing. The computer technology is blossoming. So, the DSSAT has the potential to reduce substantially the time and cost of field experimentation necessary for adequate evaluation of new cultivars and new management systems. Several crop growth and yield models built on a framework similar in structure were developed as part of the DSSAT package. The package consists of :

1. Data base management system for soil, weather, genetic coefficients, and management inputs.
2. Crop-simulation models.
3. Series of utility programs.
4. Series of weather generation programs.
5. Strategy evaluation program to evaluate options including choice of variety, planting date, plant population density, row spacing, soil type, irrigation, fertilizer application, initial conditions on yields, water stress in the vegetative or reproductive stages of development and net returns.

Weather Data for Modelling

The national meteorological organizations provide weather data for crop modelling purposes through observatories across the globe. In many European countries weather records are available for over 50 years. In crop modelling the use of meteorological data has assumed a paramount importance. There is a need for high precision and accuracy of the data. The data obtained from surface observatories has proved to be excellent. It gained the confidence of the people across the globe for decades. These data are being used daily by people from all walks of life. But, the automated stations are slowly being used. The problems are the position of sensors, mode of logging and manufacturer. Often, the calibration of different sensors is reported to be difficult. There is a huge gap between the old time surface observatories and present generation of automated stations with reference to measurement of rainfall. The principles involved in the construction and working of different sensors for measuring rainfall are not commonly followed in automated stations across the globe. The output obtained is often substantially different from surface observatories. As of now, solar radiation, temperature and precipitation are used as inputs in the DSSAT.

Weather as an Input in Models

In crop modelling weather is used as an input. The available data is ranging from one second to one month at different sites where crop-modelling work in the world is going on. Different curve fitting techniques, interpolation and extrapolation functions, etc., are being followed to use weather data in the model operation. Agrometeorological variables are especially subject to variations in space. It is reported that, as of now, anything beyond daily data proved unworthy as they are either overestimating or under estimating the yield in simulation. Stochastic weather models can be used as random number generators whose input resembles the weather data to which they have been fit. These models are convenient and computationally fast, and are useful in a number of applications where the observed climate record is inadequate with respect to length, completeness, or spatial coverage. These applications include simulation of crop growth, development and impacts of climate change. In 1995 JW Jones and Thornton described a procedure to link a third-order Markov rainfall model to interpolated monthly mean climate surfaces. The constructed surfaces were used to generate daily weather data (rainfall and solar radiation). These are being used for purposes of system characterization and to drive a wide variety of crop and live stock production and ecosystem models.

The present generation of crop simulation models particularly the DSSAT suit of models have proved their superiority over analytical, statistical, empirical, combination of two or all etc., models so far available. In the earliest crop simulation models only photosynthesis and carbon balance were simulated. Other processes such as vegetative and reproductive development, plant water balance, micro-nutrients, pest and disease, etc., are not accounted for. The statistical models use correlative approach and make large area yield prediction and only final yield data are correlated with the regional mean weather variables. This approach has slowly been replaced by the present simulation models by these DSSAT models. Similarly, mechanistic models describe the plant growth and development processes with mathematical equations. The practical utility of these models to solve agricultural problems is very limited. It is believed that many inputs like soil minerals, pests, diseases, and weeds are going to be added to the DSSAT simulation models. So, the whole soil-plant-atmosphere system is simulated.

When many inputs are added in future the models become more complex. The modelers who attempt to obtain input parameters required to add these inputs look at weather as their primary concern. They may have to adjust to the situation where they develop capsules with the scale level at which the input data on weather available.

Data Characterisation and the Application of Simulation Models and Role of Weather in Decision Making

Time span and spatial scale are two essential considerations in any agro-climatic characterization. Early methods often covered only one parameter, a derived parameter, a single application or means of presentation. Methods incorporating multiple parameters and producing multiple results were developed when understanding of crop modelling gained ground.

Decisions based solely upon mean climatic data are likely to be of limited use for atleast two reasons. The first is concerned with definition of success and the second with averaging and time scale. In planning and analyzing agricultural systems it is essential not only to consider variability, but also to think of it in terms directly relevant to components of the system. Such analyses may be relatively straightforward probabilistic analyses of particular events, such as the start of cropping seasons in West Africa and India. The aim of strategic analysis is to provide a probabilistic framework for long term decision making. This permits the rational comparison of alternative systems and answers questions as to whether a particular system would provide sufficient and stable output to justify adoption. There are well established causes and effects reported widely in the literature, meaning that the principal effects of weather on crop growth and development are well understood and are predictable. Crop simulation models can predict responses to large variations in weather but their ability to predict small fluctuations in yield is less good. At every point of application weather data are the most important input. The main goal of most applications of crop models is to predict commercial out-put (grain yield, fruits, root, biomass for fodder, etc.). In general the management applications of crop simulation models can be defined as :

1. Strategic applications (crop models are run prior to planting).
2. Practical applications (crop models are run prior to and during crop growth).
3. Forecasting applications (models are run to predict yield both prior to and during crop growth).

The crop simulation models are used in USA and in Europe by farmers, private agencies and policy makers to a greater extent for decision making. Under Indian and African climatic conditions these applications have an excellent role to play. The reasons being the dependence on monsoon rains for all agricultural operations in India and the frequent dry spells and scanty rainfall in crop growing areas in Africa. Once the arrival of monsoon is delayed the policy makers and agricultural scientists in India are under tremendous pressure. They need to go for contingency plans. These models enable to evaluate alternative management strategies, quickly, effectively and at no or low cost. To account for the interaction of the management scenarios with weather conditions and the risk associated with unpredictable weather, the simulations are conducted for at least 20-30 different weather seasons or weather years. If available, the historical weather data, and if not weather generators are used presently. The assumption is that these historical data will represent the variability of the weather conditions in future. Weather also plays a key role as input for long-term crop rotation and crop sequencing simulations.

Previously, one of the limitations of the current crop simulation models was that they can only simulate crop yield for a particular site. At this site weather (soil and management) data also must be available. It is a known fact that the weather data (and all these other details) are not available at all locations where crops are grown. To solve these problems the Geographical Information System (GIS) approach has opened up a whole field of crop modelling applications at spatial scale. From the field level for site-specific management to the regional level for productivity analysis and food security the role of GIS is going to be tremendous.

There are various methods for handling the future weather data requirements in crop modelling. The first one is to use historical weather data and to run the system for multiple years. Instead of historical weather data the generated data can also be used. If multiple years of historical or/and generated weather data are used as input a mean and associated error variable can be determined for predicted yield as well as for other predicted variables. Over time the error will become smaller, as the uncertain weather forecast data are being compared one can evaluate the risk associated with each management decision, using both mean and error values of each predicted variable.

It is expected that delivery of weather data via the world wide web as well as the operation of simulation models via the world wide web will be new agro-technologies for the near future. It seems that crop simulation models can play a critical role in crop yield forecasting applications if accurate weather information is available, both with respect to observed conditions as well as weather forecasts. During the actual growing season, the current data until the previous day are used as input. For the reminder of the season, either historical weather data or generated weather data are used. In some cases, some type of short-term or long-term weather forecasts might available, which can be used to modify the weather data, inputs to represent future conditions. Most weather forecasts are available in a qualitative format, and the crop models actually require weather inputs in a quantitative format. More work will be needed to transform the weather forecasts, both short term and long term forecasts, into a format that can be used by crop simulation models. It will be important to link the crop simulation models to the local short and long-term weather forecasts. This will improve the yield predictions and provide policy makers with advanced yield information to help manage expected famines and other associated problems.

CLIMATE CHANGE

Climate change is defined as, "Any long-term substantial deviation from present climate because of variations in weather and climatic elements".

Causes of Climate Change

1. The natural causes like changes in earth revolution, changes in area of continents, variations in solar system, etc.
2. Due to human activities the concentrations of carbon-dioxide and certain other harmful atmospheric gases have been increasing. The present level of carbon-dioxide is 325 ppm and it is expected to reach 700 ppm by the end of this century, because of the present trend of burning forests, grasslands and fossil fuels. Few models predicted an increase in average temperature of 2.3 to 4.6°C and precipitation per day from 10 to 32 per cent in India.

Green House Effect

The effect because of which the earth is warmed more than expected due to the presence of atmospheric gases like carbon-dioxide, methane and other tropospheric gases. The shortwave radiation can pass through the atmosphere easily, but, the resultant outgoing terrestrial radiation can not escape because atmosphere is opaque to this radiation and this acts to conserve heat which rises temperature.

Effects of Climate Change

1. The increased concentration of carbon-dioxide and other green house gases are expected to increase the temperature of the earth.
2. Crop production is highly dependent on variation in weather and therefore any change in global climate will have major effects on crop yields and productivity.
3. Elevated temperature and carbon-dioxide affects the biological processes like respiration, photosynthesis, plant growth, reproduction, water use, etc. In case of rice increased carbon-dioxide levels results in larger number of tillers, greater biomass and grain yield. Similarly, in groundnut increased carbon-dioxide levels results in greater biomass and pod yields.

4. However, in tropics and sub-tropics the possible increase in temperatures may offset the beneficial effects of carbon-dioxide and results in significant yield losses and water requirements.

Proper understanding of the effects of climate change helps scientists to guide farmers to make crop management decisions such as selection of crops, cultivars, sowing dates and irrigation scheduling to minimize the risks.

Role of Climate Change in Crop Modelling

In recent years there has been a growing concern that changes in climate will lead to significant damage to both market and non-market sectors. The climate change will have a negative effect in many countries. But, farmers adaptation to climate change-through changes in farming practices, cropping patterns, and use of new technologies - will help to ease the impact. A study also found that subsistence farmers adapt less to climate change than do commercial farmers. DSSAT is going to play an important role in this direction for the benefit of subsistence farmers. The variability of our climate and especially the associated weather extremes is currently one of the concerns of the scientific as well as general community. The application of crop models to study the potential impact of climate change and climate variability provides a direct link between models, agrometeorology and the concerns of the society. As climate change deals with future issues, the use of General Circulation Models (GCMs) and crop simulation models proves a more scientific approach to study the impact of climate change on agricultural production and world food security compared to surveys. It is worth trying the outputs from the GCMs to modify the weather data useful for crop modeling. The GCMs are trying various approaches to provide future climatic conditions both in absolute and relative terms. Several GCMs developed by different climate groups have several limitations. The common reason for this variation among them is large spatial scale.

Cropgro (DSSAT) is one of the first packages that modified weather simulation generators or introduced a package to evaluate the performance of models for climate change situations. Irrespective of the limitations of GCMs it would be in the larger interest of farming community of the world that these DSSAT modelers look at GCMs for more accurate and acceptable weather generators for use in models. This will help in finding solutions to crop production under climate change conditions.

The connection of the Southern Oscillation episodes with El-Nino and their impact on agricultural production will assume significance in near future. This is needed to present optimum scenarios to farmers to provide them with various options to adopt their crop management regime for the current growing season. This associated and anticipated problem in decision making may bring the DSSAT modellers and GCMs much closer.

Future Issues Related to Weather on Crop Modelling

The application of crop simulations has become more acceptable in agricultural community during last few years. For any application of a crop model, weather data is one of the key inputs. It will be critical that weather data continue to be collected for all regions where agricultural production is an economic source of income. There is an urgent need to develop standards for weather station equipment and sensors installation and maintenance. It is also important that a uniform file format is defined for storage and distribution of weather data, so that they can easily be exchanged among agrometeorologists, crop modellers and others working in climate and weather aspects across the globe. Easy access to weather data, preferably through the internet and the world wide web, will be critical for the application of crop models for yield forecasting and tactical decision making. To combat the slow progress in developing crop simulation models during the last decade it will be critical for modelling groups to improve particularly with reference to weather parameters. An open source code policy and easy exchange of crop models and modules will aid in the overall improvement of the models. It is time to see that models are operated not only by the researchers but also by farmers, consultants, extension workers and decision makers at a more rapid rate than at present. It is also important that models maintain their credibility. Validation procedures play an important role in establishing the credibility of models, improving their relevance and acceptability.

As model is a simplification and abstraction from the real system, the performance levels predicted by the model will differ form those of the real system. The question is whether the difference is so great as to be of practical importance. Any testings (subjective, statistical, etc.) can never prove that a model is valid in absolute sense, even under management and climatic conditions for which the real system has been observed. Crop simulation models will never be a substitute for experimental data collection. Field data continue to be needed for improvement. Weather and agrometeorology will be critical components in all these aspects. As our society becomes more computerized there will be more scope for the application of crop simulation models to help provide guidance in solving real-world problems related to agricultural sustainability, food security, the use of natural resources and protection of environment. Professor Gerrit Hoognboom, university of Georgia Griffin, USA, is a world famous scientists in these outstanding works.

Applications and Uses of Crop Models

1. Helps in understanding of research :

   (a) Test scientific hypothesis.

   (b) Highlight where information is missing.

2. Since, the fields become records these models help as data organization tools.

3. Integrate across disciplines.

4. Assist in genetic improvement :

   (a) Evaluate optimum genetic traits for specific environments.

   (b) Evaluate cultivar stability under long term weather.

5. Hypothesize climate change effects :
   (a) Consequences of elevated carbon-dioxide.
   (b) Changes in temperature and rainfall.
6. Assist in agronomic (cultivar) management;
   (a) Determine optimum planting date.
   (b) Determine best choice of cultivar.
   (c) Evaluate weather risk.
   (d) Determine optimum strategy for fertilizer application including investment decisions.
7. As an in-season tool helps in scheduling irrigation and application of pesticides.
8. Assist in planning to minimise erosion and leaching of agro-chemicals.
9. Yield gap analysis:
   (a) Compare actual yield to potential yield.
   (b) Attempt to account for yield reduction from insects and diseases.
   (c) To overcome the problems in obtaining potential yield.

CLIMATIC CLASSIFICATION

Temperature and precipitation play a vital role in the very existence and modification of vegetation. The factors effecting the climates over the globe are latitude, altitude, topographic features on the earth, etc. When innumerable elements are involved in governing the climate at each spot on the earth, the first step to classify the "Population" is to bring them into few groups having common characteristics. Classification brings or introduce simplicity and order in a multiplicity of individuals, which is the basis for classification of climates. If a new crop is to be introduced one should know not only the climatic conditions under which it thrives but also the climate of the area where it is to be introduced. Climatic classification also attempts to arrive at a few broad generalised climatic zones where in a more or less homogenous climate prevails.

Purpose of Climatic Classification

The purposes of climatic classification are :

1. To obtain an efficient arrangement of information in a simplified and generalised form for convenience.
2. To find the relationship among climate, vegetation and soil.

Different Climatic Classifications

For several years scientists recognised that there is a strong relationship between distribution of vegetation and climatic elements.

1. Decandole (1855) published an important monograph on the factors influencing the distribution of plant species.
2. Grisebach (1866) prepared the world map of vegetation.
3. Linsser (1867 and 1869) undertook basic studies on the response of plants to heat and the effect of precipitation on vegetative growth.
4. Koeppen's Classification :

   The most widely used system of climatic classification in its modifications is that of Wladimir Koeppen (1845-1940). He aimed for an applied scheme that would relate climate to vegetation but provide an objective, numerical definition of climate types in terms of climatic elements. Koeppen devised his first classification (1900) largely on the basis of vegetation zones and later (1918) revised it with greater attention to temperature, rainfall and their seasonal characteristics.

   The Koeppen system includes five major categories which are designated by capital letters. In order to represent the main climatic types, additional symbols are added. Except in dry climates, the second letter refers to rainfall regime, the third to temperature characteristics and the fourth to special features of the climate (Table 10.1). Several meteorologists, climatologists and geographers have made modifications of Koeppen classification. They simplified the world climatic map and redefined several climatic types.

TABLE 10.1

MAIN CLIMATIC TYPES OF KOEPPEN'S CLASSIFICATION

Climatic Group	Label	Dry Period	Degree of Drymass and Coldness
Tropical forest climate	A	F (S), W	
Dry climates	B	-	S W
Warm temperate rainy	C	F, S, W	
Cold forest climates	D	F, (S), W	
Polar climates	E	—	T F

Where

F = Precipitation well distributed throughout the year

S = Summer dry

W = Winter dry

(S) = to indicate this variation

E = This is divided into ET and EF, where

ET indicates least warmest month above 0°C and EF indicates all months below 0°C.

5. Thornthwaite classification :

(a) First, Thornthwaite (1931) classified the climates based on the "Precipitation effectiveness ratio", i.e., he divided monthly precipitation with total monthly evaporation i.e., P/E.

(b) The sum of 12 monthly P-E ratios is calculated and is designated as "P-E index".

(c) Finally, the classified the climates based on P-E indices and associated indices (Table 10.2).

TABLE 10.2

THORNTHWAITE CLASSIFICATION BASED ON P/E INDEX

Humidity proviusce	Vegetation	P/E Index
A_1 Wet	Rain forest	>128
B_1 humid	Forest	64-127
C_1 Sub-humid	Grass land	32-63
D_1 Semi-arid	Steppe	16-31
E Arid	Desert	<16

(d) He also recognised the importance of temperature and calculated temperature and efficiency T-E ratio (Table 10.3).

$$T.E = \frac{T-32}{4}$$

Where T = Temperature of the months in °F.

TABLE 10.3

THORNTWAITE CLASSIFICATION BASED ON THERMAL EFFICIENCY (T-E) INDEX

	Temperature Efficiency	T - E Index
A'	Tropical	>128
B'	Mesothermal	64-127
C'	Microthermal	32-63
D'	Traiga	16-31
E'	Tundra	1-15
F'	Frost	0

(e) In (1948) he introduced the concept of PET and also he estimated soil water balance with the help of PET and precipitation (P).

(f) He computed water balance with the following indices.

$$\text{Humidity Index (I h)} = 100 \times \frac{\text{Water surplus}}{\text{Annual P.E}}$$

$$\text{Aridity Index (I a)} = 100 \times \frac{\text{Water deficit}}{\text{Annual P.E}}$$

Combined moisture Index (Im) = Ih – 0.6 Ia.

$$\text{Im} = \text{Ih} - 0.6\ \frac{100\text{S} - 60\text{d}}{\text{PE}}$$

The positive values indicate the moist climate and negative values dry climates.

(g) Thornthwaite classified the climates into 9 climatic moisture types based on annual moisture index.

TABLE 10.4

THORNTHWAITE CLASSIFICATION BASED ON ANNUAL MOISTURE INDEX

Type	Moisture index
A per humid	100 and above
B_4 Humid	80 to 100
B3 Humid	60 to 80
B2 Humid	40 to 60
B1 Humid	20 to 40
C2 Moist sub humid	0 to 20
C1 Dry sub humid	– 33.3 to 0
D Semiarid	– 66.7 to – 33.3
E Arid	– 100 to – 66.7

6. Thornmthwaite and Mather (1955) revised the classification and introduced the soil moisture at exponential rate during the dry periods.

$$\text{St} = \text{Fc e}\ \frac{\text{APME}}{\text{Fc}}$$

Where,

St = Soil moisture storage

Fc = Field capacity

P = Precipitation

APME = Accumulated negative

e = neparian base.

7. With the modification to Thornthwaite – Mather Classification the moisture index becomes :

$$Im = In - Ia = \frac{P - PE}{PE} \times 100$$

8. Budyko (1956) delineated the climatic regimes corresponding to the radiation types, on the basis of radiational index of dryness.

   Humid month $P - PE > 0$

   Dry month $P - P E < 1$

9. Cocheme and Franquin (1967) computed monthly water availability periods.

10. (a) Papadakis (1966) proposed a classification based on seasonal temperature, seasonal distribution and availability of moisture.

    (b) Papadakis (1975) introduced the concept of monthly climate based on thermic and hydric indices.

11. Hargreaves (1971) proposed a classification based on the degree of moisture adequacy.

12. International Rice Research Institute (IRRI, 1974) classified rice growing zones based on wet, dry and intermediate months.

13. Planning Commission (1989), Government of India divided India into 15 regions on the basis of soil type, rainfall, temperature and water resources.

Agroclimatic Classification

Many of the climatic classifications have been found unsatisfactory to measure and compare the agricultural productivity of different areas, as also the relationship between climate and productivity. To judge the influence of climate on different crops and to solve several other ground level problems faced from previous climatic classifications, the new concept of agriclimatic classification was introduced later on.

Agroclimate

The combined influence of climatic elements that make possible the cultivation of crops.

Agroclimatic Index

An index which expresses the quantitative relationship between climate and agricultural production.

Agroclimatic Region

A region in which all the climatic parameters are almost homogenous as far as agricultural crops and their production is concerned.

Advantages of agroclimatic regionalisation

There are several advantages of agroclimatic regionalisation.

1. The agricultural potentiality of a region can be exploited based on soil and climatic resources.
2. The agroclimatic maps are useful for transfer of technology from one region to other.
3. The agroclimatic maps also help in introducing new crops from other regions.
4. The maps also help to develop suitable zone specific technology depending on the constraints identified.
5. To carry zone specific research.

Techniques and methods in agroclimatic regionalisation

There are two steps in delineating an agroclimatic region.

1. Quantify the main climatic characteristics.

   Example : Expressing temperature, rainfall in numerical values or indices.

2. Mapping the geographical distribution of the indices in the form of an agroclimatic region.

The weather elements are grouped into two categories based on their influence on crop growth and development.

1. Whether elements that influence vegetative phase
   - Moisture stress.
   - Solar radiation and temperature.
   - Duration of frost-free period.

2. Weather elements that influence, reproductive, maturity and other phenophases.
    - Day length.
    - Annual variation in radiation and temperature.
    - Daily range of temperature.
    - Duration of frost – free period.
    - Duration of rainy and dry seasons.

    Based on the above weather climatic elements combined indices are proposed to build the crop – climate relationship. This objective can be achieved by the evaluation of the agroclimate :
    - Where the crops originated.
    - Of other regions where the crops are grown.
    - Of regions where crops can not be grown.

Based on the above lines several methods were developed to delineate the agroclimatic regions in India.

Agroclimatic classification based on rainfall : The country is divided into 5 zones (Table 10.5).

TABLE 10.5

AGROCLIMATIC CLASSIFICATION BASED ON RAINFALL

S. No.	Zone	Rainfall (mm)	Areas in India	Crops grown
1.	Arid	<250	Desert areas	Short duration crops with supplement irrigation and fodder crops.
2.	Semi-arid	250-500	Parts of Rajasthan, Gujarat, Punjab and mid-Maharastra	Jowar, Grams, Bajra, etc.
3.	Sub-humid	500-1000	Madhaya Pradesh, Uttar Pradesh, Kartnataka, Parts of A.P. and Maharastra.	Dry paddy, vegetables, few oil seed crops, etc.
4.	Humid	1000-1500	East U.P, Karnataka, West A.P., Parts of Maharashtra	Upland paddy, vegetables, flower plants,
5.	Per-humid (Wet)	>1500	Asian Northern West Bengal	Low land paddy, etc.

In 1989, the planning commission, Government of India divided the country into 15 regions on the basis of agroclimatic factors like soil type, rainfall, temperature, water resources, etc. They are :

1. Western Himalayan region.
2. Eastern Himalayan region.
3. Lower Gangetic plains region.
4. Middle Gangetic plains region.
5. Upper Gangetic plains region.
6. Trans Gangetic plains region.
7. Eastern plateau and hills region.
8. Central plateau and hills region.
9. Western plateau and hills region.
10. Southern plateau and hills region.
11. East coast plains and ghats region.
12. West coast plains and ghats region.
13. Gujarat plains and hills region.
14. Western dry region.
15. The islands region.

Chapter - 11

Micrometeorology and Weather Modification

" The killer of the soul, whoever he may be, must enter into the plantes known as the worlds of the faithless, full darkness and ingnorance".

MICROMETEOROLOGY

Micrometeorology is defined as, "The science concerned with the detailed examination on the micro scale basis of the physical and meteorological factors that are taking place within the ecosphere". The word ecosphere means the area between just above the crop canopy and just below the root zone of a crop.

The climate of a region determines the extent of adaptability of a crop species and the weather influences its day to day growth. The crops in turn not only modify their own microclimate and weather within their canopies but also the soil underneath them due to emission of longwave radiant energy.

For general meteorological purposes the microclimate is defined as, "The climate experienced in a valley of small area when compared to that experienced over the entire mountain". But, for the crops the microclimate is defined as, "The climate near the ground in which the plants live". It differs from the macroclimate primarily in the rate at which changes occur with elevation and time.

WEATHER MODIFICATION

Microclimate of the Vegetated Surfaces

All the weather elements within the crops changes at various stages of crop growth. The differences between fallow and cropped surfaces change in terms of roughness, insolation, moisture regime, temperature, wind, humidity, etc.

1. Within the crop the profiles of solar radiation, temperature, wind speed, vapour pressure, carbon-dioxide demand depends upon the internal stand structure.

2. The profiles of these weather elements above the crop canopy appears similar to those of simple non-vegetated surface. However, they differ a lot in vegetated surfaces.

3. During the day time, canopy top absorbs or intercepts maximum insolation where maximum heating takes place. The temperature decreases both ways upwards and downwards. During night canopy emits longwave radiation and gets cooled, so that the temperature increases with height upwards and downwards within the vegetation (Figure11.1). The Co_2 profiles follows wind.

4. For wind speed profile , minimum is found in the mid to upper canopy,where foliage density is maximum, then a zone of slightly higher speeds in the more open stem layer appears and finally decreasing again to zero at the ground.

5. During the day time humidity profiles are very straight forward. Both the soil and canopy surfaces are the sources of moisture. This profile

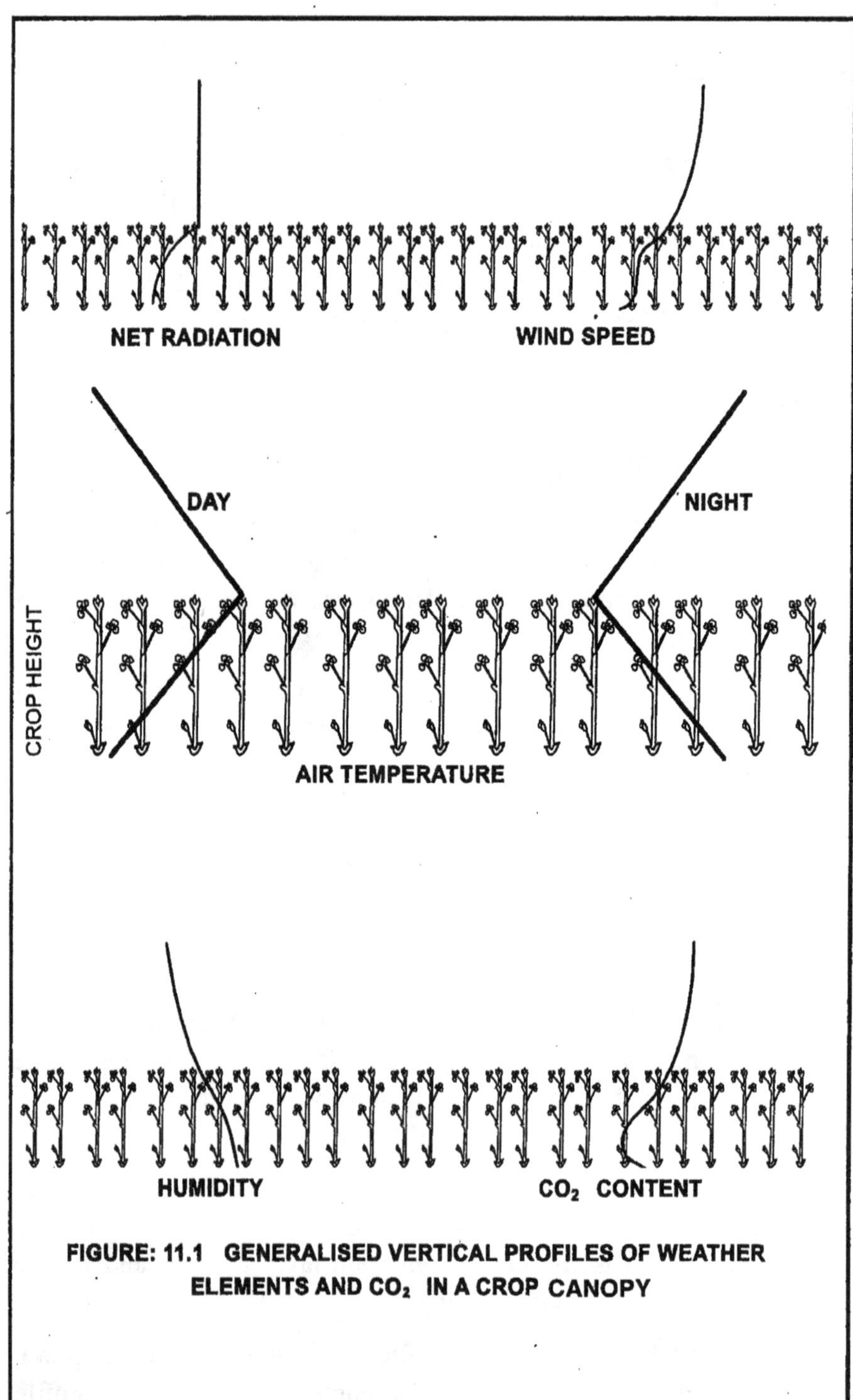

FIGURE: 11.1 GENERALISED VERTICAL PROFILES OF WEATHER ELEMENTS AND CO_2 IN A CROP CANOPY

decreases with height all the way from the soil surface up through the vegetation layer and into the atmosphere. The reduced turbulent transfer in the crop near the soil permits the vapour to accumulate.

Modification of Microclimate

Change in weather effects growth and development of plants. The extent of these effects depends upon the range or severity of change of weather. The plants can tolerate adverse weather conditions to a certain extent and beyond that limit growth and development are very much affected. Any further severity in weather ultimately results in death of plant. Repeated experience of tolerable weather for long makes the plant get adapted to such climate. However, it is very much possible to change the weather in a relatively small area to reap better harvest. Micro-weather modification is defined as, "The manipulation of weather elements at increasing or decreasing the duration, intensity, quality, etc., for desired effects at micro level of crop under investigation". The modifications to the physical environment are broadly grouped into three categories.

1. Controlling the Heat Load

(a) *Heat evasion* (in summer) : This can be achieved through shading, irrigation, etc.

(b) *Heat trapping* (in winter) : This can be achieved through planting the seeds or plants on sunny walls or slopes, erecting alternate rows of low stone walls, covering the soil between plants with white plastic sheets, etc.

2. Controlling Water Balance

(a) Increasing the water at root zone : This can be achieved through proper tillage methods.

(b) *Reducing the evapotranspiration :* This can be achieved through mulches, thinning, application of antitranspirants, etc.

3. Controlling Winds

This can achieved through shelterbelts, windbreaks, etc.

Modification at Microscale and Toposcale

Meteorological knowledge about the energy exchanges and transports at the surface has useful applications in agriculture. The changes on a small scale are relatively easy to initiate and control. Each weather element can not be modified individually by following a certain technique. If a method is followed it may result in a favourable change by modifying more than one weather element. A few examples of practical relevance are presented below.

I. Modification of Solar Radiation

Modification of radiation can be approached in two ways :

1. The manipulation of incoming radiation during the day.
2. The manipulation of outgoing radiation at night.

1. The Manipulation of Incoming Radiation During the Day

The radiation regime on a given crop field can be modified in the following ways.

A. Increasing the surface absorptive power

Darkening of the soil surface by coal dust will increase the absorptivity of incoming radiation(reduce the albedo of the surface). Therefore, the soil temperature will increase. This change may prevent the cold wind damage and prolongs the crop duration.

B. Increasing the surface reflective power of surrounding objects

The reflection by white washed walls, aluminium reflectors, etc., contributes a large amount of energy to the air, soil, vegetation, etc.

C. Increasing the exposure through site selection

With a steeper south or south-west facing slope and with an increase of width of crop rows, more radiant energy is received in a unit area at high latitudes. This can be used as an advantage depending upon the individual field problem.

D. Increasing the radiant energy by fog dissipation

Application of dry ice or silver iodide(Chapter 8) seeds clouds or fog. This in turn improves the available radiant energy.

E. Reduction of day length

(a) Artificial shading

Reduction of day length can be achieved by shading the plants in the afternoon. Tomato, soybean and other plants may flower if the day length is reduced by using cloth tents, etc., as shading material.

(b) Natural Shading

A taller and robust companion crop will provide natural shading to a shorter crop.

Example : Black pepper is grown under the shade of silver oak plants.

2. The Manipulation of Outgoing Radiation During Night

A black cloth or shade tents made of cotton reduces the effective out going radiation thereby increases temperature. Off -season vegetables can be grown successfully by using this technique.

II. Modification of Temperature

Modification of temperature is possible only if radiation is modified as detailed above. However, traditional practices like mulching, different tillage and inter cultivation practices modify the soil and ambient air temperatures.

Mulching is defined as, "Application or creation of some cover which reduces the vertical transfer of heat , water vapour, etc., benefiting the crop production".

Classification of mulches

I. Based on the Type of Material

Depending upon the type of material used mulches are divided into three categories.

1. ***Decomposing material :*** Straw of cereal crops, hay of legumes, etc.

2. ***Non- decomposing material :*** plastics, transported sand, etc.

3. ***Soil tillage :*** Harrowing, inter-cultivation, etc.

II. ***Based on the nature of material***

In this category there are two types :

1. ***Organic mulches :*** Cut grass, stems, leaves, etc.

2. ***Inorganic mulches :*** Ash, stones, plastic sheets, etc.

There are other types of mulches like *insitu* mulches(dust, weed, stubble, straw, etc.,) and imported or redeployed mulches (Aluminium foil, coal, etc).

In mulched soils daily and yearly surface temperature variations are poorly transmitted downwards. In this case the subsoil is sheltered from short term temperature excesses. When a soil is covered with a mulch the radiation can not be transmitted well into the soil and it heats up the surface. Similarly, if at night the out going surface radiation can not be replenished from below, the surface cools more. As a result, the diurnal temperature variation just above a mulch is greater than it would be on the surface of the bare soil.

Effect of Surface Geometry on Temperature

The ridges, furrows, trenches, etc., alter the amount of solar radiation striking the surface there by the temperature. The day time geometrical relationships between direct sunshine and slope and aspect of small ridges and furrows depends on their orientation with respect to slopes and position of the sun. In the early morning the tops of the ridges are colder than their bases both because of excessive sky view and because of the relatively large distance to deep soil heat. There after sunny slopes heat up appreciably, depending on orientation, also because reflected radiation may still be absorbed at the other side (radiation trapping). During the winter or dry season the north-south oriented ridges, where plants are planted on west facing slopes survived, because in the early shade they were able to make use of occasional morning dew.

Appendix - I

Table 1. Conversion Factors

In these conversion tables, all units are shown in general.

Where SI units differ from technical metric units, the conversions are given for both.

The following list details the main SI units and their symbols which are used throughout these tables.

Length

1 Km	0.62137 mile
1 m	1.09631 yd
	3.2808 ft
1 cm	0.393701 in
1 mm	0.03937 in
1 μm	39.3701 μ in
1 mile	1.60934 km
1 yd	0.9144 m
1 ft	0.3048 m
1 in	25.4 mm
1 milli-in (thou)	25.4 μm
1 μ in	0.0254 μm

Volume, capacity

1 m^3	1.30795 yd^3
1 dm^3 (litre)	0.035311 ft^3
	0.21997 imp gal
	1.7605 pint
	0.2642 US gal

Volume, capacity (Contd...)

1 cm^3 (ml)	0.06102 in^3
	0.0352 fl oz
1 litre (dm^3)	0.21997 imp gal
	1.7605 pint
1 ml (cm^3)	0.0352 fl oz
1 yd^3	0.76455 m^3
1 ft^3	28.3168 dm^3
1 in^3	16.3871 cm^3
1 imp gal	4.54609 dm^3
1 US gal	3.78541 dm^3
1 pint	0.56826 dm^3
1 fl oz	28.4131 cm^3

Area

1 km^2 (100 hectares)	247.105 acres
1 hectare	2.471 05 acres
(ha)	10000 m^2
1 m^2	1.19599 yd^2
1 cm^2	0.155 in^2
1 mm^2	0.00155 in^2
1 $mile^2$	2.58999 km^2
1 acre	4046.86 m^2
(4840 yd^2)	0.404686 ha
1 yd^2	0.836127 m^2
1 ft^2	0.092903 m^2
1 in^2	645.16 mm^2

Mass

1 tonne	1000 kg
	0.98420 ton
	2204.62 lb
1 kg	0.01968 cwt
	2.20462 lb
1 g	0.03527 oz
1 ton	1016.05 kg
	1.01605 tonne
1 cwt	50.8023 kg
1 lb	0.45359 kg
1 oz	28.349 g

Density

1 kg/m^3	1.686 lb/yd^3
	0.06243 lb/ft^3
1 g/cm^3	62.4280 lb/ft^3
1 ton/yd^3	1328.94 kg/m^3
1 lb/yd^3	0.593 kg/m^3
1 lb/ft^3	16.0185 kg/m^3
1 lb/in^3	27.6799 g/cm^3

Power

1 hp	745.700 w (J/s)
1 ft lbf/s	1.35582 w

Force

1 N	0.10197 kgf
	0.22481 lbf
1 kN	101.971 kgf
	224.809 lbf
1 kgf	9.80665 N
	2.20462 lbf
1 dyn	10^{-5} N
	0.224809 x 10^{-5} lbf
1 lbf	4.44822 N
	0.45359 kgf
1 tonf	9.96402 kN
	1016.05 kgf

Pressure, Stress

1 Pa	0.01 mbar
(N/m^2)	0.000145 lbf/in^2
1 kPa	0.01 kgf/cm^2
(kN/m^2)	10 mbar
	20.885 lbf/ft^2
	0.2953 in Hg
1 kgf/cm^2	98.0665 kPa
	14.223 lbf/in^2
1 bar	100 kPa
	14.5038 lbf/in^2

Pressure, stress (Contd ...)

1 mbar	100 pa
	2.0885 lbf/ft^2
	10^3 dyne cm^{-2}
	0.750 mm of Hg
1 atm	101.325 kPa
	14.6959 lbf/in^2
1 mm Hg	133.322 Pa
(torr)	0.01934 lbf/in^2
1 mm H_2O	9.80665 Pa
	0.001422 lbf/in^2
1 lbf/in^2	6.89476 kPa
	0.0703 kgf/cm^2
	68.9476 mbar
1 lbf/ft^2	47.8803 Pa
	0.4788 mbar
1 $tonf/ft^2$	107.252 kPa
	1.094 kgf/cm^2
1 in Hg	3.38639 kPa
	0.491 lbf/in^2
1 ft H_2O	2.98907 kPa
	0.030 kgf/cm^2
	22.3997 mm Hg

Viscosity, Dynamic

1 Pa s (N s/m^2)	0.0208854 $lbfs/ft^2$
1 cP	2.08854×10^{-5} $lbfs/ft^2$
(centipoises)	0.001 Pa s
1 lbf s/ft^2	47.8803 Pa s
1 lb/ft s	1488.16 cP
	1.48816 kg/m s

Viscosity, Kinematic

1 m^2/s	10.7639 ft^2/s
1 cSt	5.58001 in^2/h
(centistokes)	1 mm^2/s
	10^{-6} m^2/s

Viscosity, kinematic (Contd...)	
1 ft^2/h	0.092903 m^2/h
	25.8064 cSt
1 in^2/s	645.16 mm^2/s
	645.16 cSt

Energy	
1 MJ	0.27778 kWh
1 J	0.737562 ft lbf
1 kgf m	9.80665 J
	7.23301 ft lbf
1 therm	105.506 MJ
1 kWh	3.6 MJ
1 Btu	1.05506 kJ
1 Watt h/m^2	0.0860 cal cm^{-2}
	0.0860 langley
	3600 J/m^2
	0.317 BTU ft^{-2}

Table 2. SI Units and Symbols Prefix Used With Units

Multiple	Prefix	Symbol
10^{-15}	femto	f
10^{-12}	pico	p
10^{-9}	nano	n
10^{-6}	micro	μ
10^{-3}	milli	m
10^{-2}	centi	c
10^{-1}	deci	d
10	deca	da
10^2	hecto	h
10^3	kilo	k
10^6	mega	M
10^9	giga	G

Table 3. Conversion Factors (Energy)

Units and conversion factors

Gigajoule (GJ) = 10^9 joules
1 barrel oil = 6.3 GJ
Tonne wood (air dry) = 15 GJ
1 tonne oil = 44.7 GJ = 7.1 barrels
Tonne wood (bone dry) = 20 GJ
1 tonne coal = 28 GJ
Tonne agricultural residues = 13 GJ
1 tonne coal = 1.7 tonne wood = 2.3 m^3 wood
Tonne dung = 15 GJ
1 GJ = 280 kWh
Tonne charcoal = 30 GJ
1 kWh = 3.600 x 10^6 J
Tonne wood = 1.4 m^3 wood
1 GJ = 0.95 x 10^6 Btu
1 m^3 wood = 0.73 tonne
1 GJ = 0.24 x 10^6 K calories
1 m^3 wood = 10 GJ
1 calorie = 4.187 J
1 m^3 wood = 0.33 tonne coal equivalent (tce)
1 GJ = 26 litres kerosene
12 tonne wood produce 1 tonne charcoal (by earth kiln)
1 litre gasoline = 3.5 x 10^7 J

Table 4. Units and Conversion Factors for Radiation Data and Temperature

To convert values in units of the left-hand column, multiply by the relevant factors in order to obtain corresponding values in units of the top line.

A. Quautity of Radiation Per Unit Area

Units	$J\ m^{-2}$	$m\ Wh\ cm^{-2}$	$Wh\ m^{-2}$	$Cal\ cm^{-2}$
$erg\ cm^{-2}$	10^{-3}	2.78×10^{-8}	2.77×10^{-7}	2.39×10^{-8}
$J\ m^{-2}$	1	2.78×10^{-5}	2.778×10^{-4}	2.39×10^{-5}
$J\ cm^{-2}$	10^{4}	0.278	2.778	0.239
$Wh\ m^{-2}$	3.6×10^{3}	0.1	1	0.0861
$mWh\ cm^{-2}$	3.6×10^{4}	1	10	0.861
$kWh\ m^{-2}$	3.6×10^{6}	100	10^{3}	86.1
$cal\ m^{-2}$	4.19×10^{4}	1.163	11.6	1
$Btu\ ft^{-2}$	1.136×10^{4}	0.316		3.16
0.271				

B. Radiant Flux Per Unit Area

Units	$mW\ cm^{-2}$	Wm^{-2}	$Cal\ cm^{-2}\ min^{-1}$ (ly/min)
$W\ m^{-2}$	0.1	1	1.433×10^{-3}
$mW\ cm^{-2}$	1	10	0.01433
$kW\ m^{-2}$	100	10^{3}	1.433
$Cal\ cm^{-2}\ min^{-1}$	69.8	698	1
$Btu\ ft^{-2}\ h^{-1}$	0.316	3.16	4.52×10^{-3}
$Btu\ ft^{-2}\ min^{-1}$	18.9	189	0.271

Table 5. Some important units and constant values

Appendix - I (Continued)

Name & values	Units
The diametre of the Sun	1.39×10^6 KM
The distance between the Sun and the Earth Surface	1.5×10^8 KM
Temperature of the Sun	5762°K
Velocity of light	3×10^{10} cm sec^{-1}
Planck's constant	6.625×10^{-27} erg sec^{-1}
Stefan Boltzman constant	5.67×10^{-5} erg cm^{-2} sec^{-1} K^{-4}
Universal gas constant	8.314×10^7 erg $°K^{-1}$ $mole^{-1}$
	1.986 Cal $°K^{-1}$ $mole^{-1}$
Saturation vapour pressure and partial vapour pressure	mbar
Relative humidity	%
Absolute humidity	gm^{-3}
Evapotranspiration and PET	Cal cm^{-2} min; mm h^{-1}; mm day^{-1}
Flux of radiation	Cal cm^{-2} min^{-1}
Slope of SVP and temperature curve and psychrometric constant	m bar deg^{-1}

Appendix - II

A. Approximate conversion factors (light)

Day light, full sun

950 W m^{-2} = 1.36 cal cm^{-2} min^{-1} ≈ 95000 lux. Of this PAR (400-700 nm) is :
1800 μ mol photons m^{-2} s^{-1} ≈

399 W m^{-2} = 0.572 cal cm^{-2} min^{-1} = 42 % of total.
∴ W m^{-2} (total) ≃ 1.895 μ mol m^{-2} s^{-1} (PAR).

Blue sky light

72 W m^{-2} = 0.103 cal cm^{-2} min^{-1} = 9000 lux. Of this PAR is : 200 μ mol photons m^{-2} s^{-1} = 45 W m^{-2} = 0.065 cal cm^{-2} min^{-1} = 63 %
∴ 1 W m^{-2} (total) = 2.778 μ mol m^{-2} s^{-1} (PAR).

Energy flux density

1 W m^{-2} = 1 J m^{-2} s^{-1} = 2.388 x 10^{-5} cal cm^{-2} s^{-1}.

B. PPFD calculations

Watts/m^2 = 1.433 x 10^{-3} ly/min
Ly/day = cal/cm^2/day
Solar radiation 1 ly/day = 0.09 E m^{-2} d^{-1}

For Hyderabad

0.0860 ly = 3600 J m^{-2}
1 ly = 41860.465 J m^{-2} 1 MJ = 10^6 J

Conversion from μE m^{-2} s^{-1} to W m^{-2} of PAR
Divide μE m^{-2} S^{-1} by 4.6 to get W m^{-2} of PAR
1 E m^{-2} d^{-1} = 1 x 106 μE m^{-2} d^{-1}
1 day = 24 x 60 x 60 = 24 x 3600 Seconds
= 11.6 μE m^{-2} s^{-1}
= 86400 seconds

Solar radiation in ly/day to Cal cm^{-2} min^{-1} of PAR
Ly/day x 0.09 = E m^{-2} d^{-1}
E m^{-2} day-1 x 11.6 = μE m^{-2} s^{-1}

$$\mu Em^{-2}\ s^{-1} \times \frac{1}{4.6} = w\ m^{-2} \text{ of PAR}$$

W m-2 of PAR x 1.433 x 10^{-3} = Cal cm-2 min-1 or ly/min
Solar radiation in Cal cm^{-2} min^{-1} of PAR to MJ m^{-2} d^{-1} of PAR
Cal cm^{-2} min^{-1} x 41860.465 = J m^{-2}

$$J\ m^{-2} \times \frac{1}{10^6} = MJ\ m^{-2} \text{ for a minute}$$

MJ m^{-2} for a minute x 1440 = MJ m^{-2} d^{-1} of PAR
Ly/day x 0.041855 = MJ m^{-2} d^{-1} of total solar ....(1)
Ly/day x 0.0119605 = MJ m^{-2} d^{-1} of PAR band ....(2)
E m^{2} d^{-1} x 0.2178 = MJ m^2 d^{-1} of PAR band ....(3)
Ly/day x 0.09 = E m^{-2} d^{-1} of PAR ....(4)
(2)/(1) gives percentage of PAR energy (MJ m^{-2} d^{-1}) in total solar energy (MJ m^{-2} d^{-1}) which is 47 %.

C. Important abbreviations

AVHRR	Advanced Very High Resolution Radiometer (provides vertical temperature profiles).
CGIAR	Consultative Group on International Agricultural Research.
DIFAX	Digital facsimile system (a weather data transmission system).
ECMRWF	European Centre for Medium Range Weather Forecasts.
FAO	Food and Agricultural Organization.
GARP	Global Atmospheric Research Project (the overall multinational plan for a series of large field experiments).
GCM	General Circulation Model.
GMT	Greenwich Meridian Time.
GOES	Geostationary Operational Environmental Satellite
IMD	India Meteorological Department.
MONEX	Monsoon Experiment, a component of GARP.
NCMRWF	National Centre for Medium Range Weather Forecasting.
UNEP	United Nations Environmental Programme.
WCAP	World Climate Applications Programme.
WCDP	World Climate Data Programme.
WCIP	World Climate Impact Studies Programme.
WCRP	World Climate Research Programme.
WCP	World Climate Programme.

Appendix I and II are adapted from :

1. Latimer, J. Ronald. "Radiation measurement. AES. Environment. Canada. IFYGL. Technical manual series no.2, 1971.
2. World meteorological Organization (WMO). "Guide to meteorological instruments and observing practices". 3rd edition, Geneva, Switzerland, 1969.
3. Catalogue. "Weather measure, weathertronics". Division of Qualimetrics Inc., California, 1988.

References

This book is exclusively meant for beginners in this field of study and particularly to those who join in agricultural sciences with basic studies in their local languages. So, to avoid irritation and confusion of repeated references in the diagrams, tables, text, explanation, etc., all the important works utilised by the author are mentioned in this section of references only. The author whishes to inform that it is imposible to be original in an attmept of this nature and he endeavoured to acknowledge all sources of information and expressions of the original aouthors. Still, if any inadvertant omission is found the author acknowledges the same, with utmost reverence. The author expresses his thanks to all the other authors and publishers whose books he frequently consulted and referred to in this work. In addition to so many lecture notes, catalogues, cyclostyled papers, IMD manuals, monographs, instruction bulletins, ICAR publications, research notes, etc., the following books have been consulted in preparing this work.

Albright, John G. 1939. Physical Meteorology. Prentice Hall, Inc. Englewood Cliffs, N.J, 330 pp.

Brutsaert, W. 1982. Evaporation into the atmosphere. Reidel/Kluwer, Dordrecht, 299 pp.

Baldy, C., Stigter, C. J. 1997. Agrometeorology of multiple cropping in warm climates. Oxford and IBM Publ. Comp., New delhi, India, 237 pp.

Chang, Jen-Hu. 1968. Climate and Agriculture : An Ecological Survey. Aldine Publishing Company. Chicago, USA, 304 pp.

Doorenbos. J. and Pruitt,. W.D. 1984. Guidelines for predicting crop water requirements. FAO Irrigation and Drainage. Rome, Italy : Food and Agriculture Organization, 72-74 pp.

Ghadekar S. R. 2001. Meteorology. Agromet publishers, Nagpur, Maharastra, India, 251 pp.

Gupta U. S. 1982. Physiological aspects of dryland farming. Oxford and I B H publishing company, India, 322 pp.

Geiger, R., Aron, R. H., Todhunter, P. 1995. The climate near the ground (5th edition). Vieweg, Germany, 528 pp.

Geiger, R. 1965/1995. The climate near the ground (4th edition). Harvard Univ. Press, 611 pp.

Griffiths, J. F. (ed). 1994. Handbook of agricultural meteorology. Oxford University Press, United Kingdom, 320 pp.

Gordon Y. Truji, Gerrit Hoogenboom and Philip K. Thornton. 1998. Understanding options for agricultural production. Kluwer Accdemic publishers, Dordrecht, the Netherlands, 400 pp.

Howard J. Critchfield. 1998. General climatology (4th edition). Prentice-Hall of India, New Delhi, 453 pp.

Jackson, I. J. 1989. Climate, water and agriculture in the tropics (2nd edition). Longman, United Kingdom, 377 pp.

Jones, H. G. 1992. Plants and microclimate. Cambridge University Press, U.K., 428 pp.

Kaimal, J. C., Finnigan, J. J. 1994. Atmospheric boundary layer flows, their structure and measurement. Oxford University Press, 289 pp.

Lenschow, D. H. 1986. Probing the atmospheric boundary layer. American Meteorological Society, Boston, 269 pp.

Linacre, E. 1992. Climate data and resources. Routledge, New York, 366 pp.

Lowry, W. P. Lowry, P. P. 1989. Fundamentals of biometeorology : the physical environment. Peavine (Oregon, United States), 2 vols., 650 pp.

Lal D.S. 1989. Climalatology. Chaitaniya publishing company, Allhabad, India, 429 pp.

Mani, A. 1989. Development of integrated crop growth prediciton model for irrigated agriculture. M. Sc., thesis; Division of Agricultural Engineering, IARI, New Delhi.

Mavi, H.S. 1986. Introduction to agrometeorology. Oxford and IBH publishing comapany, New Delhi, India, 237 pp.

Monteith, J. L., Unsworth, M. 1990. Principles of environmental physics (2nd edition). Anrold, Unitied Kingdom, 291 pp.

Murthy, V. R. K. 1995. Practical manual on agricultural meteorology. Kalyani publishers, Ludhiana, India, 86 pp.

Murthy V. R. K. 1995. Terminology in agricultural meteorology. Srivenkateswara publishers, Hyderabad, 230 pp.

Oke, T. R. 1987. Boundary layer climates (2nd edition). Methuen/Routledge, United Kingdom, 435 pp.

Rosenberg, N. J., Blad, B. L., Verma, S. B. 1983. Microclimate : the biological environment (2nd eddition). Wiley, United States, 495 pp.

Shivakumar, M.V.K., Wallace, J.S., Renard, C. and Giroux, C. 1991. Soil water balance in Sudano-Sahelian Zone. Proceedings of the International Workshop, February 1991, Niamey, Niger. IAHS Publication no. 199, Wallingford, UK : IAHS Press, Institute of Hydrology.

Index

C

D

K

L

M

N

O

P

Q

R

S

T

www.ingramcontent.com/pod-product-compliance
Lightning Source LLC
Chambersburg PA
CBHW050523190326
41458CB00005B/1647

* 9 7 8 9 3 5 2 3 0 0 7 4 7 *